RESPONSIBLE
LIVING
IN
A
BOTTLED
WATER
WORLD

MARION
SCHAFER

Cover image © Shutterstock.com

Kendall Hunt
publishing company

www.kendallhunt.com
Send all inquiries to:
4050 Westmark Drive
Dubuque, IA 52004-1840

*I would like to dedicate this, my first book,
to my wonderful wife Jane.*

Contents

Contents

Contents

Chapter 1

Introduction

This book is primarily about responsible living in a bottled water world. Intended as a text for the Foundational Studies course on ethics and social responsibility, the focus here is on environmental concerns of packaging.

This chapter will contain elementary information about packaging, followed by a study of ethical systems and ways of viewing environmental issues in chapters 2, 3, and 4. Most people do not give much thought to packaging unless they are compelled to do so, maybe for reasons of difficulty in handling it. Otherwise, people tend to look at packages only to see the product that is inside, not the package itself. In some ways this is good, because the package exists only for the product that it contains. It is nevertheless important for businesses to think about the need for efficient and responsible packaging, considering environmental impacts of adopting or ignoring certain packaging related choices.

Packaging

Packaging has become an essential part of life today. Without modern packaging, food would spoil and be contaminated, products would be damaged while being transported and handled, many products would be difficult to store and use, and information about products would not be readily available. As true is the fact that packages and packaging have no practical purpose or value without a product that needs it.

Functions

Ideally, packages must be developed to perform certain specific functions. If a package fails to perform its functions effectively, it is obviously not a proper package for the intended product. The following are the primary functions of a package:

- Containment
- Protection/Preservation
- Information/Motivation
- Utility/Ease-of-use/Transportability

Containment is the most basic function of a package. Containment describes the ability of a package to hold a product within itself. Every package must be able to contain, or hold something to enable moving it from one place to another. The ability of a package to contain one product does not necessarily mean its ability to do so for another product. For example, a cloth bag may be perfectly capable of containing apples, but is not capable of containing water. Likewise, a cup will contain water, but will not contain compressed air. Thus it is evident that the characteristics and form of the product must be considered when determining which packaging forms and materials are suitable for each product.

Protection and preservation take the containment function to a higher level. Protection takes many forms. Protection for an automobile tire may involve adding material to locations so that the tire sidewalls do not get scuffed while being handled, stored, and transported, or blocking and bracing to hold tires in place so they do not move during transport, or fall and injure those handling them. For pharmaceutical products, protection involves several layers that seal these products in a container—maintaining their sterility, preserves the integrity of the products, and provides tamper-evidence so the purchaser can know that the product has not been altered since manufacture. In addition to protecting the product from harm, packaging may also perform the function of protecting consumers from the product, such as with poisons, pesticides, and other hazardous items. Preservation is critical for most food products to ensure that they do not spoil before consumption. Sometimes preservation can be accomplished by simply choosing certain materials, for instance in case of lettuce salad mix in a bag, plastic that maintains a certain balance between oxygen and carbon dioxide within the package, to maintain freshness.

Product information is important for many reasons. Basic information informs the consumer about the product that is contained in the package. It can be as simple as a label on a can that says "beans" or as elaborate as that of a pharmaceutical package with a label that states the brand name of the drug, the generic name, the manufacturer, the quantity/amount of drug in the package, the date of manufacture, the lot number of the batch, the last date by which the product can be safely used, list of active ingredients and amount of each, list of inactive ingredients and amount of each, recommended dosage and interval between doses, cautions for users, and a fold-out pamphlet that lists all the information about the drug to comply with FDA requirements. Motivation takes information to another level. Motivation involves making the package attract the attention of the consumer to entice a purchase. There are several things that provide motivation: type font and size, type color, background colors, pictures and drawings of the product and of other images, orientation of images and printing, shape of the package, size of the package, and many other factors that help the package stand out amidst the competition so the consumer will want to purchase this product instead of others.

Utility, which also implies ease-of-use or transportability, is the function that makes a product more useable for the consumer. Examples include indentations formed into a bottle that makes it easier to hold without slipping out of the hand, a screw-on cap or flip-top cap that allows the consumer to reseal the bottle if the product is not used at one time, a handle that makes it easier to carry or pour from heavy containers, a pour spout on a carton or jug, and other such features. These features are all built into the package so that the product is more user-friendly.

It is important to understand that it is possible to have packages that include all four of these functions, but not all packages are required to perform all four. It is critical that every package contains its product, even if that is all it does. For example, a pitcher may be used as a package that will hold liquid and allow it to be transported from one place to another. For a consumer who needs more than mere containment, protection may be added to the packaging function. A sealable plastic bag can add a level of protection for a sandwich so that it does not dry out while being stored or transported. It can also help in preserving the sandwich by keeping out germs and other harmful things. Date of production and type of sandwich printed on the bag serves the function of information. If this bag can be opened and then resealed, the criterion of ease-of-use is fulfilled.

Definition

At this point, a definition of packaging is appropriate. A proper package must perform its intended functions throughout its useful lifetime. Thus, the following is offered as a definition of proper packaging:

> Packaging is a system that provides containment of a product, protection of a product, and from a product, information about the product, and features that provide ease of use and transportablilty for a product, from the point of manufacture—or the initial placement of the product into the packaging system—to the point of final use—or the final separation of the product and the packaging system. (MDSPackagingConsultants.com)

Significantly, the above definition includes the four functions explained above, while adding the parameters for how long the packages are expected to perform those functions. If the package fails to perform its functions throughout the entire time for which it is needed, it has failed as a package and must be changed.

Materials

There are many materials used in packaging systems. Commonly used materials for packaging may be divided into some basic categories: a) Materials made from renewable item such as paper, paperboard, and corrugated paperboard made from plant and tree fibers, b) Materials made from mined minerals, such as steel made from iron ore, aluminum made from bauxite ore, and glass made from silica sand, c) Materials made from non-renewable resources, such as plastics made from petroleum and natural gas. Additionally, there are now plastics made from renewable resources like corn and soy.

Disposal

Packaging always has a limited lifetime. Since the usefulness of any package is tied to a product, when it is no longer needed for the product, it is necessary to make a decision about what to do with the package. The options available for facilitating this decision follow along with the list that is widely known as the "Three Rs," which are reduce, reuse, and recycle. Most people seem to focus only on the last one—recycle. It is important to know what these three terms mean in order to understand why they are stated in the order mentioned.

To "reduce" means to use less packaging, or to use less packaging materials. This sounds simple on the surface—just use less—but it is not actually so. The amount of packaging and packaging materials needed to efficiently perform the necessary functions for the product depends on the form and nature of the product.

Questions

1. What is packaging?
2. What makes a package good, in your opinion?
3. What are some examples of poor packages, in your opinion?
4. What more would you like to know about packaging that you do not understand now?

Chapter 2

Ethics

This chapter introduces the reader to ethics in engineering. Whether or not you are interested in engineering, these concepts will be helpful. As you read, please make note of the things that fit into your area of interest, and see what you can learn from the rest.

Michael R. Williamson is Assistant Professor in the Department of Applied Engineering and Technology Management at Indiana State University. He previously worked for the Illinois Department of Transportation. He has an interest in professional ethics.

Ethics for Engineers

By Michael R. Williamson

Within societies, ethical beliefs may differ from person to person. There may be various reasons for this and we will discuss some examples in the following section. The ethics of an individual can have a long and lasting impact on society because it is often related to choices that they make in everyday life.

Job Influence of Engineers

A person's career can have a definite impact on their ethical beliefs, based on the sworn duties they take an oath to follow (American Society of Civil Engineers, 2015). A good example of strong ethics are those pledged by civil engineers when they take professional oath to faithfully take on the obligation to serve the needs of the society by providing safe infrastructures that meet the demands of the local economy.

Engineers must recognize that the daily decisions they make will directly impact the safety of the common public. For example, the structures that engineers design must be strong enough to support all necessary loadings without failing. If a structure such as a building or a bridge fails, people's lives will be in danger. Engineers must have strong ethics that would prevent them from taking shortcuts that could impact the safety of the public, even if such practices were common in other industries to save costs.

In case of drinking water as well, civil engineers are entrusted with the health and welfare of the public. Some of these engineers are responsible for design and operation of water treatment plants. Great care must be taken to preserve public health, as the slightest disregard of ethics could have a catastrophic effect on public health, leading to multitudes of people getting sick or even dying.

Sustainability is also of importance and considered a prime ethical value that engineers should adopt. All projects should be designed with the intention of avoiding harm to the environment and providing sustainable solutions to problems. Unnecessary depletion of materials should not be allowed and reuse of materials on site should be encouraged. For example, when repaving a roadway, the rock needed—which is already in place—could be reused.

Additionally, engineers should only preform work within their areas of expertise and honor the integrity of the profession (American Society of Civil Engineers, 2015). Engineers as well as people in other professions must uphold the honor of their profession to keep public trust intact. There are many ways of keeping the environment safe and preventing unnecessary pollution. Keeping the trust of the general public can only be done by making judicious decisions backed by real data, taking all possible consequences into consideration.

Sustainability

Overconsumption of resources has been a common cause in the collapse of societies throughout the history of mankind (Graedel & Allenby, Industrial Ecology and Sustainable Engineering, 2010). Great civilizations across the world have been lost, as can be seen in the ruins of abandoned cities like the Angkor Wat Temple complex in Cambodia, and the many Mayan cities in Mexico. Other Native American villages located in the United States once housed a dense population. The once prosperous Native American city located at Cahokia Mounds was one of the largest cities in North America before the arrival of European immigrants. It has been proven that the city at Cahokia Mounds possessed all the characteristics of a complex society. Large plazas where citizens could gather and trade goods, time keeping devices, permanent structures, and evidence of religious symbols have all been found in the ruins of this complex hierarchical society. Citizens of Cahokia probably had a social structure similar to a modern one, thriving on trade and diverse types of employment. Because of this stability the population began to grow. Food supplies were plentiful due to the city's location along the Mississippi river; rich soil provided an array of plants for consumption—including corn—one of the predominate sources of food for the Cahokians. Evidence suggests that the corn was dried and stored for winter and a surplus was there to fall back on in case of crop failures. Hunters from the city made use of the abundant game and fishermen took advantage of the Mississippi river and its tributaries' plentiful aquatic life. The city grew from only a few thousand to an estimated 40,000 in its peak, with several smaller cities located in the peripheral areas providing additional resources to the large urban center (United Nations Educational, Scientific and Cultural Organization, 2015). Supporting so many people evidently required a large amount of resources and was dependent heavily on weather patterns and dedication of the citizens. The city was abandoned soon after a peak in population and eventually left in ruins. The great mounds that once housed the chief can be seen in figure 2.1.

So what caused this once flourishing city to collapse? Some believe that overconsumption of resources in a focused area was most likely the cause. Unlike in current times, the city would have had limited ability to move in resources from other areas in a timely manner to continually support life, but even if the Cahokian's had the ability to reach out further and obtain more resources from other areas perhaps the city would have grown even larger and consumed more resources until even those were also depleted. The collapse of the Cahokians would have just been delayed but not avoided altogether.

A study of today's complex societies makes evident the similarities between ourselves and the Cahokian civilization in regards to consumption of resources. Of course, we have advanced technology that allows

Figure 2.1 Monks Mound at Cahokia
Source: WikiMedia Commons, http://commons.wikimedia.org/wiki/File:Monks_Mound_Cahokia_3978a.jpg

us to produce superior products and transport them over great distances; however, at some point, we too will inevitably reach the limits of consumable resources available. What will happen then and when will this catastrophic event occur?

Quantifying sustainability is a meaningful way of understanding the impact of our practices on earth. Using what is sometimes referred to as the "master equation," as seen in equation 1 below, we can establish a quantitative interpretation of it (Graedel & Allenby, Industiral Ecology and Sustainable Engineeirng, 2010). As is evident, population, resource use per person, and the impact of resource use—all have a significant impact on our environment.

$$\text{Overall Environmental Impact} = \\ \text{Population} \times \frac{\text{Resource Use}}{\text{Person}} \times \frac{\text{Environmental Impact}}{\text{Unit of Resource Use}} \quad (1)$$

Population is an easily quantifiable measure that can be directly taken from census data in first world nations and estimated with some degree of accuracy in rural areas or third world nations. Resource use

11

per person is quantifiable as well, which can be estimated by surveys and by knowing the overall resource consumption of a population in a particular geographic area. The part that becomes difficult is estimating the environmental impact of some resource uses. Some resource impacts are established though advanced studies through universities or other laboratory environments that offer a non-biased view on resource consumption. Many resource impacts are unknown and require more studies, making it difficult to calculate the master equation accurately.

Carrying Capacity

Eventually every society will reach its carrying capacity, and only those who take responsibility of the inevitable will survive. Let us consider a simple example of capacity with only one resource. Corn is produced widely in Midwestern United States and supports many forms of life, including humans, both for direct consumption and for feeding animals that are later to be consumed by humans. When weather patterns cause crop shortages, the impact is controlled by price increase, causing a person to consume less or rely on another product. This is a good short term control and effective in the long run as well, as it resists reaching capacity. At some point the capacity of our corn production will be reached and then, shortages will not be controlled by price increase; people will simply have to go without. Of course, in reality, the shortages would be much more expansive than just corn and the animals that rely on it. Once humans reach the carrying capacity of the earth, countless changes will have to occur or civilization as we know it will end.

Some scientists claim that the population that earth can support is somewhere around 10 billion, based on given present conditions of available resources and a current population that is estimated to be 7 billion (Wikimedia Commons, 2015). In figure 2.2, you will notice the extreme change in population growth after 1950. While there are many factors that can be linked to the sharp increase in population, one fact that is predominant is the fact that people are living longer than ever before due to the advances in the medical field. The blue curve represents the current growth rate, which has reached nearly 7 billion. Take a look at the three possible curves past our current date, each representing possible trends in population growth. The green curve represents a situation when a peak is reached just over 8 billion, causing concern. The population is projected to decrease after the peak—by low birth rates or as consequence of some catastrophic event such as lack of resources.

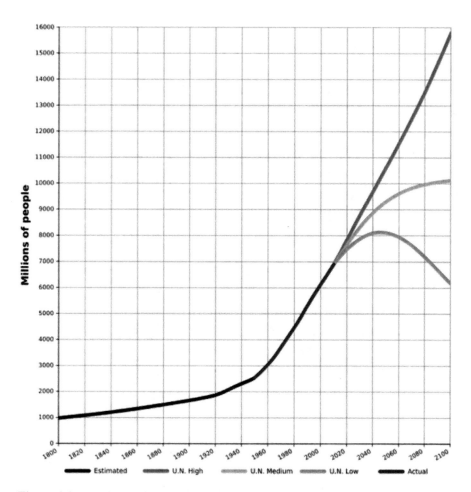

Figure 2.2 Population Growth Trends
Source: http://en.wikipedia.org/wiki/Malthusian_catastrophe#/media/File:World-Population-1800-2100.svg

In this case, a correction is made in time to prevent capacity from being reached. The yellow curve represents a situation when we level out our growth at capacity. In this situation, we would be using all possible resources, and as a result even the smallest reduction in crop production would have devastating effects. Finally, we have the third possibility, depicted by the red curve, to consider. It shows an unchanged population growth curve. If we continue at the current rate, the capacity will certainly be met or surpassed at some point. All possible resources will be consumed faster than they can be replaced, causing a collapse in societies around the world.

Other models on sustainability reflect a different pattern where the carrying capacity of the earth is constantly fluctuating based on environmental changes that may affect crops and other natural resources. Adjustments in population are made in order to compensate for the current carrying capacity. Another model predicts that the carrying capacity will decrease proportionately with an increase in population. As population rises, resources are consumed at a faster rate, decreasing the carrying capacity. Regardless of which model we choose to follow, these are predictions and the precise effect is unknown.

Transportation

An important topic in any society is that of transportation, which has a direct impact on a society's ability to progress. In the United States of America many transportation options are available, ranging from personal vehicles to mass transit, of several types. While in developing or underdeveloped nations transportation may be restricted to only a few ways, usually in the form of personal bikes and some mass transit.

Our choices regarding the mode of transportation that we use have a direct impact on the environment in the form of resource consumption and pollution. In the United States, most people feel the need to own their own vehicle for personal transportation and given the layout of most communities, having access to a personal vehicle may be vital in obtaining a job or even for shopping for groceries. Many urban areas, such as the city of Chicago, have acknowledged this problem and have taken steps to reduce the impact that transportation has on the environment. There are many mode of transportation available for use in the city of Chicago. While personal transportation in the form of automobiles is still a predominant choice for commuters, many residents do not own a car either because they avail of other available modes or because the cost of owning a car is too high. Other choices include walking and biking, which are made convenient for users with sidewalk designs that allow safe crossing of major roadways, dedicated bike lanes and bike sharing programs make biking to work possible, and an array of mass transit including trains, buses, and water taxis. Mass transit must keep to a tight schedule to ensure that riders get to their destinations in a timely manner. A major factor in mode choice is reliability. For example, if you have to be a work at 8:00 AM, you need transportation that will get you there on time. For everyday purposes, people will prefer to take a reliable mode of transport. One problem in having a complex transportation system such as in the city of Chicago is the consumption of many resources. Many people have to commute each day from suburban areas to the city center for work, resulting in a high pressure on the transportation system. According to the annual Urban Mobility Report, the Chicago roadway network costs the average commuter nearly US$1200 per year, ranking it the 5th highest in the United States resulting in over 50 hours of delay per year per commuter (Schrank, Eisele, & Lomax, 2012). In terms of fossil fuel consumption, 127,016,000 gallons are used each year just by personal vehicles.

A problem that exists in every major urban area is that people tend to work in the urban center and live in the surrounding areas. This necessitates daily commuting, causing unnecessary traffic, and resulting in unnecessary consumption of resources and increased amount of pollution. This trend is evident in most urban areas in the United States as well as in other countries. As more and more houses are built by clearing farm or woodland, more fuel is consumed in transporting people to and from desired destination, causing pollution rates to flourish as a result.

The city of Toronto is a pioneer in reversing this trend of urban sprawl, after decades of suffering the consequent adverse effects. Toronto is building up corridors through the urban center that will attract people to live, work, and shop within walking distance. The idea is to build up key arterials through the city and provide easy access for walking, biking and mass transit along the key routes. Mass transit is available in the form of streetcars and buses with very short wait times, a key factor in deciding to take mass transit for trips. People in Toronto can live the good life without the expense of owning a car, because they can avail the various modes of transport, facilitating movement in, out and through the city. A picture showing transit options can be seen in figures 2.3 and 2.4.

Figure 2.3 Multimodal transportation in Toronto
Source: Dr. Michael R. Williamson

Figure 2.4 Bike lane
Source: Wikimedia Commons

Other cities, including Chicago, are following this trend in an attempt to increase the quality of living in urban areas and prevent unnecessary consumption of resources.

So what causes this urban sprawl resulting in unnecessary consumption of resources? Much is dependent on the desire to live in a better environment, free of pollution and crime. Crime is heavy in areas in cities that are not properly maintained. As houses and other properties decay, so do the property values resulting in a lower quality of life. Families typically leave urban areas for suburban areas to escape crime, ending up consuming more resources and contributing to an increased amount of pollution.

Transit Stops

Even though the harmful effects of pollution are widely known, many people still decide to drive their own vehicles instead of taking mass transit. Many urbanized areas have greatly improved mass transit systems, making it a convenient choice for commuters. Shelters have been placed at many transit stops to provide a comfortable place for riders to wait. Even wi-fi is a feature in many public transit systems now, allowing riders to use their mobile devices on transit or sometimes at stops. Continual information is given to riders regarding wait times for the next arriving transit vehicles, reducing passenger's frustration regarding uncertain timings or long waits and helping them plan their trips accordingly. Many transit systems have employed advanced technology to attract more riders. Trip planning apps can now be used to quickly identify the quickest transit options from one's current location to a desired destination. The Chicago Transit Agency (CTA) provides quick access to maps showing how to get to the nearest transit stop, transit routes, schedules showing the arrival times for each transit stop, alerts providing information on crime or delay, transit trackers showing the location of each transit vehicle in real time and fare information showing the cost for a particular trip (Chicago Transit Authority, 2015). All such features are aimed at attracting more commuters to use mass transit instead of personal vehicles.

Additional transit choices in the city of Chicago are light rail, extended bus service, water taxis, and one of the largest cab networks in North America. It is possible to take transit from the outlying areas of the city to anywhere in the downtown areas. Bike lanes have been put throughout the city of Chicago, which can be easily accessed from residential areas to places of work or business. Rent-a-bike stations have been placed at key points, enabling people to rent a bike at one station and return it to another station of their choice. Yet in spite of such convenient, user-friendly options, many people still make the choice of driving their own vehicles.

So what are the reasons for so many people still not availing of public transit? Safety seems to be a key factor in the deciding on a mode of transport. Many feel unsafe waiting at transit stops or riding in a vehicle with other people whom they do not know. Trip times also have an influence. In areas with limited service, taking the bus or a train may not be the most convenient option. People feel more comfortable knowing they can make a trip anytime they choose to, without having to wait.

Recycling

While depletion of natural resources will ultimately lead to the demise of society, precautions can be taken to deter the inevitable and extend sustainability. While products that use less material are certainly preferable, incorporation of recycling into the manufacturing process can further extend our limited and renewable resources.

A great example that everyone is familiar with is the recycling of aluminum cans. Recycling of aluminum cans has a high participation rate; most households are willing to separate aluminum cans from other trash due to the high payoff rate. However, other materials that have little or no value are not as readily recycled. Another common factor influencing recycling practices is the fact that smaller cities often do not have the financial means to separate recyclables from trash and recycle them. Several large cities have realized the need to enact recycling programs that require residents to separate certain types of trash for recycling for the purpose of reducing landfills consumption, lowering waste disposal costs, reducing pollution rates and consumption of raw resources. Some cities still produce massive amounts of wastes that are trucked to rural areas for disposal.

A good approach in regards to encouraging recycling is to provide an easy way for citizens to recycle. The city of Chicago provides recycling bins to citizens that are picked up at regular intervals using a single stream recycling system (City of Chicago, 2015). Single stream recycling is a process where all recyclables are mixed together in one bin and separated at the recycling facility. This process requires a more complex facility at a higher cost, but comes with several benefits. Single stream recycling does not put pressure on citizens to separate different types of trash. For example, some more common materials such as newspapers, plastics, glass, and aluminum would each require their own bin within each household, taking up space. Also, the hassle of constantly having to sort produces within the home may deter many citizens from participating in recycling programs. The main advantages of single stream are that a) more people are willing to participate because of the easier task of not having to separate materials within the household, b) less space is needed for collection containers within the home because only one bin is required, c) collection costs are reduced because collection crews can move faster and utilize single bin trucks, and d) new materials can be easily collected without changes to the collection system. An example of the bins used in Chicago can be seen in figure 2.5.

In regards to communities taking a proactive approach when it comes to environmental issues, the city of San Francisco has been taking astounding steps using advanced technology in order to reduce its environmental footprint. More communities should follow San Francisco's efforts to reduce their environmental footprints, as it gives cities better appeal when they are seen as "smart cities." Smart cities strive for efficiency in mobility, construction, energy, and transportation all around making them a greener community. The individuals who live in these cities are more likely to live a greener lifestyle since they are surrounded by all suitable amenities, rather than having to go out of their way to obtain something.

Solar powered bus stops offer a great convenience for riders since they are given an estimated arrival time rather than left wondering when the bus is supposed to arrive or if they have missed a specific one. Free Wi-Fi at bus stops adds an extra incentive for riders. This ensures that are occupied while they wait and this might encourage them to choose

Figure 2.5 Recycling Bins Chicago
Source: Arvell Dorsey Jr. https://www.flickr.com/photos/chicagofd1996/8558923740/

the bus over driving themselves. If more communities adopted these features, there would be a higher percentage of bus riders than drivers. Only when the alternatives offered are clearly better than what people have generally been getting, will people be motivated enough to adopt a practice.

Recycling can be done on more than just common household trash. Of course, industrial wastes are typically recycled whenever possible, but something that you may not be aware of is the recycling that occurs with a nation's infrastructure. Massive amounts of rock used in making our roadways, bridges and buildings are consumed on a daily basis to make life as we know it possible. Those rocks are used in making concrete and asphalt where they are held together by a binder. In concrete the binder is Portland cement while in asphalt the binder is a bituminous (i.e., oil based) material. Once constructed, the binder in either cannot be reused. However, the rock known as aggregate can be reused for several times. Essentially, the base building block can be reused on other projects or to reuse the existing one. In typical repaving project on our interstate system, machines are used to grind off the top layer of asphalt where the small rocks are separated from the binder, followed by recoating the rocks with new binder before they are replaced on the roadway. Unless additional thickness is being added to the pavement, no new rock is required. Similar principles are applied when buildings or bridges are reconstructed, greatly reducing the environmental impact by reducing resource consumption.

Some countries have more stringent recycling programs where citizens are required to participate properly or pay heavy fine. Japan requires its citizens to separate recyclable materials into different bins, each product having an identifiable marking making it easier for people to know what bin to place an item in (All Recycling Facts, 2015). All manufactures of packaging and containers are required by law to place markings on their products including plastic, paper, cardboard, steel, aluminum and polyethylene terephthalate (PET) bottles. Residents are required to separate their trash into as many as 44 bins. Another law that makes Japan different from the United States is that all appliances are required to be recycled at the cost of the customer.

When comparing the amount of waste produced per country the United State is found to be far above most other countries with the average person producing 1600 pounds per year (Facts and Details, 2015). Citizens of Japan only produce an average of 875 pounds per year while poorer countries produce lesser. For example, a Mexican citizen produces an average of 600 pound per year. In developed nations there is an abundance of products citizens can choose to purchase, which explains the difference between the United States and Japan while in third world nations such as Mexico, citizens cannot afford costly items in packaging and reuse all possible available resources including packaging materials.

While there are many products that can be recycled, let us again focus on aluminum cans, which are more commonly recycled than other items, to make a comparison between countries. In the United States, aluminum cans are not required to be recycled in all areas, so the driving force is many areas is the monetary value of doing so (The Aluminum Association, 2015). As a result the current recycling rate in the United States is 66.7%, which is an historic high. While bins are often available for citizens to place cans in, some choose not to recycle. Many citizens recycle at home due to the can's value, but again not all choose to recycle. In Japan the aluminum can recycling rate was last reported to be 94.7%, one of the world's highest, while cost is again a driving factor here, with citizens requiring to recycle, causing the high rate of recycling (Aluminum for Future Generations, 2015). Third world countries have other reasons for recycling. One that surpasses all others in the recycling of aluminum cans is Brazil, with a 98.2% recycling rate. There cans are recycled due to the lack of other usable resources and the monetary value associated with it.

Life Cycle Analysis

An important concept to understand is that of "Life Cycle Analysis." Life cycle analysis involves the complex analysis of products from raw resources to disposal. Let us analyze the life cycle of a common soda can. There are eleven stages in the life of an aluminum can beginning with mineral extraction and ending with re-melting of used cans (Belinda, 2006).

The first stage involves bauxite mining, which involves the use of heavy equipment that consumes large amounts of fossil fuels to operate. The equipment life also needs to be considered in the life cycle of the aluminum can. The depleted hours in the life of the equipment in the extraction of bauxite can very well be calculated and included in our analysis. The second stage involves the extraction of alumina from the bauxite ore. Here a chemical process is used that involves heating the material to a temperature between 140 and 240 degrees Celsius depending on the quality of the ore. Pure alumina is produced by separating all other minerals. Again, large amounts of energy are consumed in the form of fossil fuels. So we again need to consider the life of the equipment used in the second stage.

The third stage is smelting, where the alumina is again heated and transformed into aluminum. It is common for 16 kWh of electricity to be consumed to produce one kilogram of aluminum from alumina. As before, we not only need to consider the electrical energy used in the production of aluminum can but the hours of life taken from the equipment in the production. In stages 5 and 6, which involve the fabrication and filling of the aluminum, energy is consumed again and life is taken from the equipment in the production.

Stage 7 consists of the product use by the consumer, followed by the re-melting of the aluminum, which is also included in this stage, resulting in high energy use with the cans being heated to 780 degrees Celsius to remove paints that have been applied to the aluminum. The transportation of the product from place to place during the production and distribution—to the vendor, to the consumer's home or place of use—and ending with the transportation of the used aluminum can to be recycled and the life thereby taken from all modes of transportation must also all be considered during the life cycle analysis.

The final four stages (8 to 11) involve the collection, sorting, re-melting, and transportation of the aluminum can, where it is re-used for more products. As is evident, the life cycle analysis of an aluminum can is fairly complicated, with much energy consumed to produce a final product, broken in to several stages of the production process. Complex models have been created to analyze similar products allowing for an in-depth understanding of our resource consumption.

The Cost of Convenience

Packaging has made products more convenient to the consumer. As we as a society progress, so does the complexity of our goods, for several reasons. In ancient times packaging was limited to mainly bulk storage units. Typically, people had reusable vessels they took to the market to purchase goods and stored them within the same vessel within their homes until use. This was practical due to the fact that people could easily travel some short distance to the market place for most of their goods. Goods shipped from longer distances would have used bulk storage containers that were reusable, such as wagons or cargo bays on ships, where they were made available to the consumer. Issues with this packaging included susceptibility to diseases and undesirable display of the product. The prevalence of diseases attributed to the development of more advanced packaging that could not be easily penetrated by pests. Packaging is a form of safety in terms of public health. Even today, first world nations have less diseases as a result of more advanced packaging, whereas third world nations have more diseases from the lack there of.

As society progressed it became inconvenient to carry vessels to the market. Products began to be shipped over longer distances and it is usually safer to ship them prepackaged for health and safety reasons. In today's society people are focused more on convenience rather than seeking safety from packaging. Take into consideration the common convenience store station, where many items are on display in all types of packaging. The attractive produces are ready to go; the consumer just makes a selection and takes it, to open later as convenient. Items in bulk bins would require one to do more work and take more time to make a selection or in some cases have a worker prepare something. For these reasons, packaging has really taken effect in our society. Most customers would frown on a bulk bin of candy bars or other food items that anyone could touch. Imagine finger prints from others all over your food of choice.

Packaging also plays a role in a person's selection of certain products. People tend to be attracted to shiny things. Because of this effect on people's choice, manufactures have become very competitive for attracting the consumer's eyes. Unfortunately, attracting business in this way has led to an increase in unnecessary packaging, leading to unnecessary consumption of resources.

According to a recent study on the effect of packaging on consumers, it was observed that a key to marketing for any business is in packaging (Lane, 2015). Packaging is a way to promote a product above the rest, by attracting the consumers' eye through the use of color schemes and bigger and better designed packages. Consumers think they are getting more when a product package is bigger and brighter. The security of packages was also found to have an effect on consumer choice (Lane, 2015). Consumers were found to favor packaging that would protect the product during shipping to prevent costly returns, which in turn requires more packaging material. However, consumers were also found to favor biodegradable packaging material or products made from recycled products.

Basic Understanding of Packaging by University Students

Students at a university typically have the following knowledge about packaging and the environment. While they have been exposed to environmental issues related to specific packaging, they often do not understand the complexity of the packaging industry. The general understanding is for what is good and what is bad for the environment. Students overwhelmingly believe that cardboard products are good for the environment due to the wide recyclability of the product. Students also feel that recyclable materials are good for the environment without any consideration for the resource depletion or other environmental effects they may cause.

One particular product that students feel is bad for the environment is plastic, due to the publicity of the garbage island in the Pacific ocean and the adverse effects on wildlife. Many people are now envisioning plastics as bad for the environment, while other products such as the aluminum can are seen as good by students who feel that when recycled, the can has an overall positive effect.

Resource Consumption

All resources are rather limited and can be depleted if not managed correctly. We need to consider two main types of resources: first those that are renewable and those that are not. Renewable resources can be regenerated if managed properly. A good example would be trees. As long as trees are replanted after harvest and managed to ensure healthy growth, we can continue to reuse trees for paper products. However if we diminish our resource of trees without replacement we will eventually deplete this resource. Other resources are limited, including precious metals. We have a limited quantity that can be obtained for use, so recycling is key to prevent depletion of the resource. Some cities have begun to mine landfills to obtain metals and other recyclable resources.

References

All Recycling Facts. (2015, May 19). *Japan's recycling pictures.* Retrieved from All Recycling Facts: http://www.all-recycling-facts .com/recycling-pictures.html

Aluminum for Future Generations. (2015, May 20). *Japan.* Retrieved from Aluminum for Future Generations: http://recycling.world-aluminium.org/regional-reports/japan.html

American Society of Civil Engineers. (2015, May 13). *ASCE.* Retrieved from http://www.asce.org/code-of-ethics/

Belinda, H. M. (2006). *Analysis of the recycling method for aluminum soda cans.* University of Southern Queensland.

Chicago Transit Authority. (2015, May 13). *Trip planners*. Retrieved from Chicago Transit Authority: http://www.transitchicago.com/travel_information/trip_planner.aspx

City of Chicago. (2015, April 22). *Recycling and waste management.* Retrieved from The City of Chicago's Official Site: http://www.cityofchicago.org/city/en/depts/streets/supp_info/recycling1/what_is_single_streamrecycling.html

Facts and Details. (2015, May 19). *Recycling in Japan.* Retrieved from Trash and Garbage in Japan: http://factsanddetails.com/japan/cat26/sub162/item869.html

Graedel, T., & Allenby, B. (2010). *Industrial ecology and sustainable engineering.* Upper Saddle River, NJ: Prentice Hall.

Jr., A. D. (2013, March 11). *Chicago recycling bins.* Retrieved from Flickr: https://www.flickr.com/photos/chicagofd1996/8558923740/

Lane, T. (2015). *Small business.* Retrieved 04 27, 2015, from Small Business: http://smallbusiness.chron.com/packaging-affect-consumers-70612.html

Schrank, D., Eisele, B., & Lomax, T. (2012). *Urban mobility Report.* College Station, Texas: Texas A&M Transportation Institute.

The Aluminum Association. (2015, May 20). *Aluminum can recycling holds at historically high levels* - Retrieved from The Aluminum Association: http://www.aluminum.org/news/aluminum-can-recycling-holds-historically-high-levels

United Nations Educational, Scientific and Cultural Organization. (2015, May 13). *Cahokia Mounds state historic site.* Retrieved from http://whc.unesco.org/en/list/198

Wikimedia Commons. (2015). *World population 1800–2100.* Wikimedia Commons.

Questions

1. Explain which ethical issues from this chapter fit into your area of interest.
2. What can you learn from those things that do not fit your current needs?
3. How are ethical issues for engineers different from those in other professions?
4. What will be an important ethical issue in your profession?
5. What is the most important thing you learned from this chapter?

Chapter 3

Religion, part a

Introduction

Religions have traditionally played an important role in shaping the sense of ethics of followers. Whether religious principles informed and directed ethical stands, or whether ethical stands were actually the basis for the development of religions is a matter of debate, and beyond the scope of this book. This chapter will introduce some of the basic stands taken by religions and let the reader decide what to believe.

This chapter discusses the three largest religions in the Western world—Judaism, Christianity, and Islam—all of which derive from a common root. These are the religions that grew out of the person of Abraham, and thus called the Abrahamic religions.

Abraham was a man who lived thousands of years ago in the middle-east. He believed that God called him to leave his parents and travel to a place that God would show him. That place is what is now known as Palestine and Israel. Abraham and his wife Sarah lived in that area for many years and amassed a sizeable fortune, but they had no children to whom they could pass on their wealth. In their old age, God promised Abraham that he would have children—as many as the grains of sand. Abraham and Sarah thought this was impossible, so Abraham begot a son, Ishmael, with Sarah's servant. Then several years later, Sarah gave birth to a son, Isaac. This is where history splits.

The Jews believe that because he was born of Sarah, Abraham's wife, Isaac is the first-born son who inherited the blessings of God, thus the heirs of Abraham are those born of Isaac. The Christian religion, which developed out of the Jewish religion, follows that same logic. The followers of Islam, however, argue that Ishmael was the first-born son of Abraham, and thus the true descendants of the first-born are those born of Ishmael. Despite this split, all three groups revere Abraham as an important figure in the founding of these religions. For this reason, their basic ethical views on environmental issues are quite similar.

Christian and Jewish Views

The Jewish and Christian views on environmental issues are, on the surface, a simple one. Their fundamental perception of environment and nature is based on the statements in the Book of Genesis in the Bible:

In the beginning God created the heaven and the earth. And the earth was without form, and void; and darkness *was* upon the face of the deep. And the Spirit of God moved upon the face of the waters. And God said, Let there be light: and there was light. And God saw the light, that *it was* good: and God divided the light from the darkness. And God called the light Day, and the darkness he called Night. And the evening and the morning were the first day.

And God said, Let there be a firmament in the midst of the waters, and let it divide the waters from the waters. And God made the firmament, and divided the waters which *were* under the firmament from the waters which *were* above the firmament: and it was so. And God called the firmament Heaven. And the evening and the morning were the second day.

And God said, Let the waters under the heaven be gathered together unto one place, and let the dry *land* appear: and it was so. And God called the dry *land* Earth; and the gathering together of the waters called he Seas: and God saw that *it was* good.

And God said, Let the earth bring forth grass, the herb yielding seed, *and* the fruit tree yielding fruit after his kind, whose seed *is* in itself, upon the earth: and it was so. And the earth brought forth grass, *and* herb yielding seed after his kind, and the tree yielding fruit, whose seed *was* in itself, after his kind: and God saw that *it was* good. And the evening and the morning were the third day.

And God said, Let there be lights in the firmament of the heaven to divide the day from the night; and let them be for signs, and for seasons, and for days, and years: And let them be for lights in the firmament of the heaven to give light upon the earth: and it was so. And God made two great lights; the greater light to rule the day, and the lesser light to rule the night: *he made* the stars also. And God set them in the firmament of the heaven to give light upon the earth. And to rule over the day and over the night, and to divide the light from the darkness: and God saw that *it was* good. And the evening and the morning were the fourth day.

And God said, Let the waters bring forth abundantly the moving creature that hath life, and fowl *that* may fly above the earth in the open firmament of heaven. And God created great whales, and every living creature that moveth, which the waters brought forth abundantly, after their kind, and every winged fowl after his kind: and God saw that *it was* good. And God blessed them, saying, Be fruitful, and multiply, and fill the waters in the seas, and let fowl multiply in the earth. And the evening and the morning were the fifth day.

And God said, Let the earth bring forth the living creature after his kind, cattle, and creeping thing, and beast of the earth after his kind: and it was so. And God made the beast of the earth after his kind, and cattle after their kind, and every thing that creepeth upon the earth after his kind: and God saw that *it was* good.

33

And God said, Let us make man in our image, after our likeness: and let them have dominion over the fish of the sea, and over the fowl of the air, and over the cattle, and over all the earth, and over every creeping thing that creepeth upon the earth.

So God created man in his *own* image, in the image of God created he him; male and female created he them.

And God blessed them, and God said unto them, Be fruitful, and multiply, and replenish the earth, and subdue it: and have dominion over the fish of the sea, and over the fowl of the air, and over every living thing that moveth upon the earth.

And God said, Behold, I have given you every herb bearing seed, which is upon the face of all the earth, and every tree, in the which is the fruit of a tree yielding seed; to you it shall be for meat. And to every beast of the earth, and to every fowl of the air, and to every thing that creepeth upon the earth, wherein there is life, I have given every green herb for meat: and it was so. And God saw every thing that he had made, and, behold, it was very good. And the evening and the morning were the sixth day. (Genesis 1, KJV)

The Torah, which is composed of the first five books of the Bible, has the same creation story as does the Christian Bible. The translation may be slightly different, but the meaning is the same. Thus, the Jewish views about the environment, based on the idea that God created everything and humans were put on earth to care for all that, are the same as those of the Christians.

Naturally, there are probably as many different representations and interpretations of specific details as there are people, but the basics are still the same, underlying these differences. Two individuals—sharing the belief that God created the earth and everything in it and has made humans responsible for caring for His creation—may very well have opposing views on how to carry out that mandate. Person A might believe that we should not kill animals, since we are supposed to care for them. Person B might believe that God created some animals for humans to eat, and therefore, killing them for food is not against the mandate of care. Who is right—Person A, Person B, or both? This is a disagreement that we will not solve here, so we just have to decide what we prefer to believe and act accordingly.

Islam

The following was written by a follower of Islam, with an emphasis on pollution rather than general ethical views on environmental issues, but should provide a sound basis for understanding the Islamist view.

Ethical views on Pollution According to Muslims

by Bouchaib Afrej

Environmental conservation is an issue of concern in many parts around the world. In the contemporary times, human activities have influenced and affected the natural balance of the environment in many negative ways. The restoration of balance in the natural world requires the interactions of human beings and the environment. Muslims account for at least 20% of the population of the world. Some nations in the world are predominantly Muslim with the result that the environmental degradation also affects the Muslim population. In many ways, capitalism can be blamed for the exploitation of the environment. Individuals exploit the natural resources in a bid to eke a means of livelihood for themselves and their families. However, according to Islamic beliefs, everything in the world that is a creation of God is beautiful and wondrous and hence should be protected (Quran: TaHa (20), ayahs 53–54). Keeping this in mind, this paper seeks to present the Muslim perspective on the environment, its pollution and environmental conservation.

It is the firm belief of the Muslim creed that all creation in the universe has value, is meaningful and purposeful as Allah created it in his wisdom (Quran: Surat ad-Dukhan (44), ayahs 38–39.). Various sections of the Islamic Holy book are cited to prove their cases and arguments towards handling of the environment. As a result, the Muslims feel obligated by their faith, if nothing else, to conserve the environment as instructed. Muslims also have a very strong belief in the teachings of their Prophet Muhammad, who taught that the dearest creatures to God are those that take care of other creatures (Hadith related by al-Bayhaqi in *Shu'ab al-Iman* on the authority of Anas). From the Quran and the teachings of the prophet, the Muslims believe that they should maintain a relationship with the universe and all that it contains in a way that promotes the sustainable utilization of resources, as well as the progress of mankind as he seeks to improve his life. Additionally, Muslims believe that taking care of the environment is a way of obeying the teachings of their faith and that they shall be rewarded by Allah in the fullness of time.

In short, the Islam teaches the use of the environmental resources in a responsible way that ensures sustainable development of the human race. The Islamic religion is against exploitation of natural resources for economic gain, disregarding care of the environment or not considering the effects and consequences of human activities (Foltz, Denny & Baharuddin, 2003). To the Muslims, conserving the natural environment and its resources while also using it responsibly for purposes of development and the well-being of other creatures brings a promise of a blessing from the Quran. Imam Bukhari states that, "If a Muslim plants a tree or sows seeds, and then a bird, or a person or an animal eats from it, it is regarded as a charitable gift (sadaqah) for him." Islam, through the Quran, also allows land reclamation and encourages farming and reaping harvest from such lands (Athar related by Yahya ibn Adam al-Qurashi). The preservation and management of the environment given by Allah is therefore a requirement for the faithful, and not simply a social responsibility.

Natural Resources

Natural resources are elements that are freely available for human use and exploitation, having no human involvement in their existence. Such elements include air, water, soil, minerals, forests, oceans, seas, rivers and animals. Muslims believe that during the period of creation, Allah created the entire natural environment and allowed human beings to use it (Finnegan, 2005). Muslims, as well as Christians believe that God handed over the earth and all its contents to humans for dominion. The Quran states that "It is He who has appointed you viceroys in the earth . . . that He may try you in what He has given you." (Quran: Sura Al-'An`ām (6), 165). Therefore, it is only natural that such a big privilege comes with an equally big responsibility, that of using resources wisely.

1. Water

Water is the basic requirement for survival, for both plants and animals. As a natural resource, it is also the most widely used and the most crucial element for survival of life. Without water, no plant, animal or human would be able to exist in the world (Quran: Surat al-Anbiya' (21), ayah 30). In the Quran, it is stated clearly that it is God who sends down water from the sky for the benefit of mankind (Quran: Surat al-Baqarah (2), ayah 164.). According to Muslims, water, apart from being the basic unit of survival, is also an important part of Muslim religion. The other purpose of water is for purification and ablution (Ozdemir, 2003). Water helps to clean dirt from the body or clothes and at the mosque, water is used as a way of purifying the body of impurities so that one can enter into the temple pure and clean. According to the Quran, rain is sent by Allah to also cleanse human beings (Quran: Surat al-Anfal (8), ayah 11). Muslims also believe that water should be available to all people without discrimination or monopoly. The Prophet Muhammad instructs Muslims to always share water, pasture, and fire (Hadith related by Abu-Dawud, Ibn Majah, and al-Khallal). Muslims believe that the use of water should be moderate and regulated to avoid wastage. Water sources should also be protected since they are also the habitat of animals in the seas and oceans and other aquatic areas.

2. Air

Just like water, air is of crucial importance to the survival of life. Animals and human beings rely on breathing air to survive. Blowing wind results in important processes in agriculture such as pollination. The air also has religious importance to the Muslims. The Quran mentions Allah sending fertilizing winds (Quran: Surat al-Hijr (15), ayah 22). These winds signify abundant production of food and success in agricultural activities. As a result of the predominant importance of air in the life of mankind, it needs to be kept from pollution so as to avoid problems related to the respiratory system (Wersal, 1995). Pollution of the air is seen by Muslims as contravening the wisdom of God in creation.

3. Soil and Land

Land and soil are important to human beings in several ways. The land is the dwelling place of most land creatures, including humans. Soil is precious as it supports the growth of plants for human and animal consumption. The Quran emphasizes that the land is the habitat of living creatures (Quran: Surat ar-Rahman (55), ayah 10) and also for all terrestrial beings (Quran: Surat Nuh (71), ayahs 17–18). Vegetation grows in the soil, which is then consumed by both animals and plants. Pollution of the soil and the land occurs in various ways and some are due to ignorance (Rice, 2006). For instance, the constant use of agrochemicals on soil can cause pollution of soil, making the land infertile. Unsuitable farming practices lead to soil erosion and flooding. Soil degradation is also a result of mining, deforestation, and monocropping, among others (Mian, Khan & Rahman, 2013).

4. Plants and Animals

Animals and plants are also crucial for the survival of human beings. Basically, all human food is either from plants or from animals and their products. On the other hand, many animals depend entirely on plants for food. Hence, without a proper balance between animals and plants, human existence is at risk of starvation and extinction (Kula, 2001). In the Islamic religion, they believe that God allows the rain to shower on the seed planted on the earth so that it can grow and flourish for the benefit of mankind and terrestrial animals (Quran: Surat 'Abasa (80), ayahs 24–32). Plants are not only important for providing food, but also for providing nutrients to the soil through the process of decomposition. Plants also counter soil erosion—by either water or wind—by holding soil particles firmly by their roots (Al-Damkhi, 2008). Animals also help the plants by providing nutritious content through their excretion. Plants and animals are very valuable to the ecosystem as their parts can also be used for manufacturing various items such as wax, oil, perfume, medicines, fuel, timber and fiber, to name but a few (Rathore & Nollet, 2012).

Taking care of animal and plant life is not just beneficial to the existence of the human race, but a prerogative. Muslims believe that they have a responsibility to take care of animals and plants, showing mercy to them so that mercy in turn shall be shown to them (Hadith related by Abu-Dawud and at-Tirmidhi on the authority of 'Abd-Allah ibn 'Amr). They also hold the belief that allowing an animal to die out of hunger or thirst is punishable by the fires of hell by God (Hadith of sound authority) Muslims therefore advocate for responsibility and moderation when fishing or hunting for the purposes of food but prohibits hunting for sporting activities (Hadith of sound authority). The Islamic law states that all animals have legal rights that can be enforced by the office of the hisbah.

Islam and Environmental Conservation

Islamic law aims at advancing the welfare of all creatures in their present state and even in the hereafter. According to the Islamic law, the term "all creatures" refers to all animals in the sea or on land as well as all human beings around the world. In this regard, the Muslim community is dedicated to ensuring the welfare of all animals in all aspects of development and progress. Islamic law recognizes individual responsibility in taking care of the environment as a religious duty. Muslims believe that they are responsible for their own lives and the decisions that they make, for which they will be answerable to God on the Day of Judgment.

Environmental degradation as a result of ignorance of Quranic law is a common problem. Some Muslims do not know what Allah expects of them in regard to the environment and as a result they injudiciously exploit the environment. The Quran clearly states, "O children of Adam! . . . eat and drink: but waste not by excess, for Allah loves not the wasters." (Quran: Surah Al-'A`rāf (7), ayah 31). More education on Islamic instructions for environmental care is needed to ensure that more individuals are engaging in efforts aimed at conservation of the environment. In this way, more individuals will be able to take up their proper responsibility in caring for the environment.

Based on Islamic law, God is the sole owner of the earth (Abu 'l-Faraj 'Abd ar-Rahman ibn Rajab, in *al-Qawa 'id)*. The Quran and Islamic teachings agree that human beings are just mere custodians of natural resources, which should be used responsibly and for purposes that are divine and pure. Despite the right to own property in Islamic law, the use of property can be restricted. Muslims also believe in pollution of the self. This involves pollution from eating too much food. Pollution of the self can also come from filling the mind with many negative things or ideas that are prohibited by Allah. Islamic teaching states that certain things are forbidden, in order to teach the value of self-control. Some of the forbidden things that Muslims believe have the ability to pollute the body include pork, carrion, (Quran Surah Al-Mā'idah (5), ayah 4) and alcohol (Quran Surah Al-Mā'idah (5), ayah 90).

Pollution of the environment is strongly forbidden by the Islamic religious teachings. Muslims believe that it is the responsibility of the human beings to take good care of the environment as it is essentially a command from God. Taking care of the environment is a way of pleasing God and acquiring blessings. In contrast, misuse of the environment in the Muslim religion is regarded as disrespect for the will of God, hence a very serious sin. One of the ways of taking care of the environment is through feeding and watering animals, using appropriate farming techniques to avoid soil erosion and flooding as well as avoiding food wastage. In doing this, Muslims believe that God will reward them for being faithful custodians of His earth.

References

Al-Damkhi, A. M. (2008). Environmental ethics in Islam: Principles, violations, and future perspectives. *International Journal of Environmental Studies, 65*(1), 11–31.

Ali, M. M. (2011). *Holy Quran.* Retrieved from eBookIt.com.

Armstrong, S. J., & Botzler, R. G. (Eds.). (1993). *Environmental ethics: Divergence and Convergence.* New York, NY: McGraw-Hill, 275–276.

Finnegan, E. (2005). Islam and ecology. *Environmental Ethics, 27*(1), 101–104.

Foltz, R. C., Denny, F. M., & Baharuddin, A. (Eds.). (2003). *Islam and ecology: A bestowed trust.* Cambridge, MA: Centre for the Study of World Religions, Harvard Divinity School.

Haq, S. N. (2001). Islam and ecology: Toward retrieval and reconstruction. *Daedalus, 130*(4), 141–177.

Khalid, F. M. (2002). Islam and the environment. In P. Timmerman (Ed.), "Social and Economic Dimensions of Global Environmental Change." (Volume, 5, 332–339). Chichester: John Wiley and Sons, Ltd.

Kula, E. (2001). Islam and environmental conservation. *Environmental Conservation, 28*(01), 1–9.

Llewellyn, O. A. (2003). The basis for a discipline of Islamic environmental law. In R. C. Foltz, F. M. Denny, & A. Baharuddin (Eds.), *Islam and Ecology: A bestowed trust* (185–247). Cambridge, MA: Centre for the Study of World Religions, Harvard Divinity School.

Mangunjaya, F. M., & McKay, J. E. (2012). Reviving an Islamic approach for environmental conservation in Indonesia. *Worldviews: Global Religions, Culture, and Ecology, 16*(3), 286–305.

Mian, H. S., Khan, J., & Rahman, A. (2013). Environmental ethics of Islam. *Journal of Culture, Society and Development, 1*, 69–74.

Ozdemir, I. (2003). Toward an understanding of environmental ethics from a Qur'anic perspective. In R. C. Foltz, F. M. Denny, & A. Baharuddin (Eds.), *Islam and Ecology: A bestowed trust* (3–38).

Cambridge, MA: Centre for the Study of World Religions, Harvard Divinity School.

Rathore, H. S., & Nollet, L. M. (Eds.). (2012). *Pesticides: Evaluation of environmental pollution*. CRC Press.

Rice, G. (2006). Pro-environmental behavior in Egypt: Is there a role for Islamic environmental ethics? *Journal of Business Ethics*, *65*(4), 373–390.

Wersal, L. (1995). Islam and environmental ethics: Tradition responds to contemporary challenges. *Zygon®*, *30*(3), 451–459.

Questions

Following are some questions you should be able to answer after reading this chapter. Since this is not an exhaustive study of the view of Abrahamic religions, it would be helpful for you to research other sources to add to your knowledge and help you develop your ethical views.

1. Why do the Abrahamic religions have similar views on environmental issues?
2. What are the views on environmental issues that the Abrahamic religions have in common?
3. Explain the ethical views of one of the Abrahamic religions (not your own religion), and provide a critique of those views.
4. Share your own views on environmental issues.
5. Identify one or more questions that you would like to get answers for, about one of these religions.

Chapter 4

Religion, part b

In chapter 3, you read about views of the three Abrahamic religions on environmental issues. In this chapter, we look at the same issues, but through the lens of Native Americans, and non-religious people. To add to your knowledge and broaden your understanding, you are encouraged to do some research on the environmental views of other religions, such as Hinduism, Buddhism, Taoism, and any others you can find information on.

Native Americans' Views of the Environment

Mostafa Madani and Marion D. Schafer, Ph.D.

The Native Americans are considered the pioneers of environmental conservation initiatives. According to Barry and Frankland (2014), the Native American culture teaches about the existence and importance of harmony between humans and the natural environment. Native Americans view man as an integral part of the society to whom the responsibility of protecting and conserving the environment is bestowed. This worldview by the Native Americans contributed to the inception of geology, which was a new area of study in environmental conservation.

Environmental ethics encompasses a series of interventions that inform humans on how they ought to behave or act towards the natural life. The core values of Native Americans are in agreement with environmental conservation strategies, at the same time adhering to a rich cultural heritage that champion for the conservation of resources.

Chief Seattle's Vision of Environmental Conservation

A study by E.J. Bruce (2010) reveals that Native Americans followed a set of unique ethics, which are highly attributed to Chief Seattle. The values propagated ideas that made people conscious of the need to conserve the environment. He further notes that Seattle gave the analogy of society and man where he likens the relationship of man and the environment to that of blood family members.

Traditionally, Native Americans have primarily been farmers. They mostly practiced shift cultivation in which they carried out their farming activities on a virgin piece of land. Once the minerals were depleted, they abandoned the land and moved on to another fertile piece of land. This system led to massive deforestation and damage to the natural resources. Chief Seattle emphasized the need for the Native Americans to appreciate farming practices that could champion harmony between man and his environment for posterity. His views that every part of the land has to be treated as a sacred place for his people cultivated the spirit of environmental conservation. Native Americans were not only concerned with the exploitation of resources, but also with ways of engaging in best practices that would lead to protection of the environment and natural resources (Hoogland, 2013).

Besides farming, Native Americans have also been engaged in hunting. Men hunted for game meat to supplement their diet. Despite the fact that there was plenty of meat, Native Americans only choose specific parts of it, leaving the rest to rot. For instance, Boylan (2013) noted that Native Americans who hunted buffalos that inhabited cliffs and mountainous regions consumed only particular parts of the game meat. The rest would be left for scavengers, or left to decay. Chipewyan Native Americans, on the other hand, slaughtered muskox and caribou. An interesting aspect of their eating habits was that they would take a few tongues for consumption then leave the rest of the meat to decompose.

Efforts of conserving soil through the addition of organic matter can be traced back to the times of the Native Americans. This community understood the essence of humus in the conservation of the soil. Decaying organic matter from animal tissues could become humus. The manure helped in holding the soil particles together, thus preventing land degradation processes like soil erosion and denudation (Meier, Ute, Krause, & Merle, 2014).

Initially, the demand for hides, game meat, and fur was relatively small owing to the dispersed settlement pattern of the Native American population. B. Bunya (2011) further notes that the small number of Native Americans meant a moderate demand for wildlife compared to the abundant supply in the wilderness. However, depletion of game meat resulted from the tragedy of the "commons scenario." Vivian (2009) describes this phenomenon as a condition that happens when no one can claim ownership of a particular natural resource and that anyone can access it at will. The wildlife in this scenario was the commons. No one can claim ownership of animals in the bush until they have been killed. If you left an animal in the forest, it was likely that someone else would kill it.

The fact that no one had direct control over how exploitation of wildlife was to be conducted created a condition that threatened the population of wildlife. There was no incentive to responsibly take care of these natural resources. Animals like mastodon, mammoth, saber-toothed cat and ground sloth became extinct because of overexploitation by skilled hunters. The situation worsened with the coming of European settlers and colonial masters who had the capitalistic mindset that paid no regard to environmental conservation. Settlers created an instant demand for wildlife products. The Europeans wanted fur and hides for their textile industries. This development compelled the Native Americans to trap some of these indigenous animals in exchange for goods like clothes, firearms, beads, and knives. There existed neither structures nor legal frameworks that regulated the hunting. The process led to the extinction of a large number of fur bearing animals (Yael and Seideman, 2009).

A study by Serena and Warms (2009) was cynical of the idea that Native Americans lived harmoniously with the nature. They argued that claiming that Native Americans carried out their activities without causing any harm to the natural environment was synonymous with saying that they never touched anything that existed in the physical environment, a fact that would contradict history. The history of a community, according to Merchant (2013), is defined by the manner in which members of that particular community interact with the natural environment in shaping their future.

Although Native Americans often manipulated their environment to bring a desired change that could guarantee them existence and sustainability, their activities caused minimal impact on the environment compared to those of their colonial masters from Europe. The Europeans would exploit any resource they were able to lay their hands on, for the sake of their industries and for maximization of profits. The Europeans considered America a wilderness and the idea of conserving land or natural resources in such an environment was not only absurd, but also impractical. The Native Americans however shaped their perception of ecosystem profoundly since their culture worked in unison with a belief in posterity, and practiced accordingly (Bruce, 2010).

Incentives for Environmental Conservation

Native Americans were conscious of the need to develop a culture where members of the community took responsibility in matters concerning environmental protection. This drive existed despite exceptional conditions that caused a "tragedy of commons scenario,d where no one bothered about being accountable for what happens to the natural environment. The Native American community in America emphasized that a persony took responsibility in matters conceritual values and personal ethics (Meier et al. 2013). However, values and ethics were supposed to work hand-in-hand with communal and private ownership of properties and rights. The rights defined frameworks that clearly explained conditions one had to fulfill before exploiting any resource in the environment. These rights also provided structures used to identify and reward people who were engaged in good environmental stewardship.

There are many similarities in the way Native Americans traditionally organized their activities in comparison to the modern age. First, the modern government has the responsibility of protecting the natural environment through the establishment of legislation that governs exploitation of resources. For Native Americans, the word nation referred to various Native American tribes even though they did not work through organized formal structures of a government. The majority of these tribes were composed of small groups with independent administrative structures. The only time that these tribes came together was during major celebrations.

Native Americans did not have a formal written language. Bunya (2011) argues that this barred them from developing a formal legal framework that could be useful in communicating their ethics on issues concerning matters of the environment to members of the community.

Rights to Resources

The existence of institutions that clearly defined who has a right to fishing, hunting territories, land and personal property made Native American amass a lot of wealth. Hoogland (2013) reiterated that the prevalence of structures in Native American history that guided people on judiciously exploiting particular resources, created the modern property ownership rights—a fact that conditioned interference in the natural ecosystem by human beings.

Rights to land and water

Land tenure system varied significantly among Native Americans. Land ownership was entirely communal. Individualistic ownership was highly discouraged by the community. There was a scarcity of land owing to the degree of private land ownership, a fact that made it difficult to enact legislation that governed the rights to land ownership. Those people who owned land made massive investments in the area since it was easy to demarcate. However, ownership of land in the Native American community was a reserve of the clan or the family, unlike the current individual land ownership. For instance, family members among Mohican Native Americans had the hereditary privilege of using well-structured tracts of land along riverbeds for cultivation. Europeans, however, acknowledge this system of land ownership. Whenever they wanted to buy land, they had to seek the consent of family heads (Boylan, 2013).

Those Native Americans who lived in the southwest engaged in settlement farming, and private land ownership was shared here.

Every family harvested their produce and kept it in their personal stores. Contributions to the public store were voluntary, and the chief who was in charge of public needs used the produce here in the event of an acute food shortage. In the eastern part of this country, the chief assigned each family some land, since there were large tracts of land. The chief used to oversee the cultivation process and allowed each family to harvest their plot.

The Pueblo Native Americans in Colorado developed land ownership rights, which reflected their aspirations to conserve the environment. The Hopi sub-clan took advantage of the periodic floods in Colorado basin during summer by erecting stone barriers, which was crucial in checking the speed of the water (Vivian, 2009). The strategy enhanced retention of moisture in the soil while at the same time protecting the crops from floods. This technique formed the basis of modern flood control and of irrigation as a strategy to conserve soil.

Rights to Hunting

In areas where Native Americans primarily depended on hunting and fishing, there was a need to control the access to hunting and fishing grounds to prevent overexploitation. This was in response to the "tragedy of commons," a situation in which no one cared about nature, as everyone was for himself. The customary laws that governed hunting and fishing grounds were expressed in spiritual and religious spheres, unlike the conventional methods. These methods were relayed in science and technology. Nonetheless, the rules were effective in conserving the environment at that time. The strategies gave rise to modern aquatic and wildlife conservation plans (Bunya, 2011).

Establishment of hunting territories was a common scenario among Native Americans who lived in the northern part of the country. Individuals were allowed to hunt at a given place for a stipulated period and move to the next place. Rules guiding the hunting activities were established.

51

Hoogland (2013) reports that whenever a hunter marked a territory, no other hunter would willingly or knowingly encroach on that region for whatever reason, for generations. In some areas, families owned hunting grounds and the search for game meat would be done on a rotational basis. This practice helped to avoid depletion of the resources in a particular region. The ownership of such territories could be passed down through generations. The main essence of this ownership was to guarantee a constant supply of game meat and vegetables to the family while at the same time to protect the environment from external invasions by trespassers.

Paiute Native Americans who lived in Owen Valleys of California also protected their gathering, hunting, and fishing territories from trespass. Boylan (2013) reports that the community nominated individuals who would take charge of securing territorial borders. The protection of these hunting and gathering grounds came with benefits to environmental conservation, since natural tree lines, ridges, mountains, and rivers bound the restricted areas. These are significant resources in support of biodiversity. No group was supposed to encroach on territorial grounds of boundaries belonging to a neighboring clan unless in exceptional cases like possibility of starvation. However, the chief still had the overall say, as he was the one allocating the hunting territories, and he always advocated for the collective management of each area. Hunting was also controlled by norms and customs of harvest. The Native American communities had particular headmen who recommended and approved of best hunting times based on their experience in hunting and matters related to it.

Fishing Rights

Native Americans living in the Pacific Northwest had established effective fishing rights. Some of the fish that they captured included the salmons in the oceans and spawn in fresh water bodies. They used weirs, fish wheels and other appliances at shoals and falls where fish channelled naturally (Meier et al., 2013).

The fishing technology that the natives adopted was so effective that fishing threatened to deplete aquatic resources. The Native Americans realized the need to allow some fish to escape to encourage breeding. This aim was achieved through the practice that encouraged people to remove weirs from water, enabling some fish to escape upstream, once they had caught enough for consumption. The fish that escaped could lay eggs, which guaranteed the community a steady supply of fish in the future. Modern fish conservation strategies—like the use of standard-ized nets that capture only big fish and allow the small fish to go back to water—were borrowed from Native Americans. In addition, restrictions on fishing for some months, to allow fish to breed and grow, dates back to the time of Native Americans as well.

The Haida Native Americans mostly consumed salmons. The clan members restricted the access to fishing grounds to persons who were strictly members of their family. They excluded members of other fam-ilies from accessing their fishing territories. Those found trespassing were severely punished. *Yitsati*, the head of the house, had the final say on issues regarding fish exploitation. He was empowered to de-cide on when to fish, levels of harvesting, and the best fishing methods. Although it is not clear exactly how powerful Yitsati was, a study by Yael and Seideman (2009) reveals that there was an adequate and steady population and supply of salmon for a long period—because a result of the rules enforced by these community leaders. Eventually, the popula-tion of the fish reduced tremendously because of indiscriminate fishing by Europeans who exploited the fish for commercial purposes and with the intention of reaping maximum monetary profit.

Ownership of Private Items

Ownership of land and other natural resources varied, and so did own-ership of personal items like clothes, women-owned utensils, and houses. Women for instance, gathered hides, scraped, and turned them, in preparation for a ceremony for sewing the hides by members of the

53

community. Natural resources could be shared freely, but once a task was complete, the final product belonged to the family members who initiated the work, and not the community.

The culture of Native Americans institutionalized the concept of private ownership. Due to scarcity of resources, community members had to walk for long distances to get hard stones that they used to make arrows. This shortage led to the individualization of weapons and their properties. The use of horses also revolutionized the transport sector, whereby the horses could be used in chasing buffalo when hunting. This made the price of the horse go up, compared to other commodities (Bunya, 2011).

Incentives

The right to property played an integral part in encouraging environmental conservation in the culture of Native Americans. However, when an activity was undertaken communally, the society stipulated mechanisms to recognize individuals who exhibited exceptional abilities in the task. For instance, during the hunting of buffalo, an individual whose arrow killed the animal could have the privilege of taking the skin and other special parts of the game meat. This explains how Native Americans appreciated the use of incentives in developing institutions that fostered good relations between human beings and natural resources. Well-defined property rights encouraged community members to conserve scarce resources. Spirituality and ethics guided the manner in which an individual could relate to his environment.

Conclusion

From the discussion, it is evident that Native Americans were the pioneers of environmental conservation initiatives. The act of living in harmony with nature is deeply rooted in the Native American culture and religious beliefs. It emphasizes the need to conserve the environment for posterity. The need to have a safe environment prompted Native Americans to establish various initiatives for conserving the ecosystem. These consequently gave rise to modern conservation structures. There is a crucial need in modern societies to creating institutions that will inform the state on how resources may be judiciously exploited, and to establish frameworks that recognize and reward individuals who take leading roles in conserving the environment.

The arrival of Europeans with their extreme capitalist ideologies ruined the gains that had been realized by Native Americans over years of living in harmony with nature. The colonial masters exploited natural resources without considering a need to protect the environment, since they viewed the region as a wilderness.

References

Barry, J. & Frankland, E. G. (2014). *International encyclopedia of environmental politics*. New York: Routledge.

Boylan, M. (2013). Environmental ethics. New York: John Wiley & Sons.

Bruce, E.J. (2010). *Native Americans today: A biographic dictionary*. London: ABC-CLIO.

Bunya, B. (2011). Environmental crisis of epistemology? Working for sustainable knowledge and environmental justice. London: Morgan Publishing.

Hoogland, J. (2013). *Conservation of the black-tailed prairie dog: Saving North Americaca western grasslands*. New York: Island Press.

Meier, J., Ute, H., Krause, K., & Merle, P. (2014). *Urban lighting, light pollution and society*. New York: Routledge.

Merchant, C. (2013). *American environmental history: An introduction*. New York: Columbia University Press.

Serena, N. & Warms, R. (2011). *Cultural counts: A concise introduction to cultural anthropology.* New York: Cengage Learning.

Vivian, T. (2009). *Garbage In, Garbage Out: Solving the problems with long-distance trash transport.* Virginia: University of Virginia Press.

Yael, C. & Seideman, D. (2009). *Water pollution.* Chicago: Infobase Publishing.

Questions on Native Americans' View on Environmental Issues

1. In your view, what is harmony between humans and the natural environment?
2. How did Native Americans practice conservation of resources?
3. Discuss the impacts on the environment if American agriculture were to practice shift cultivation as the Native Americans did.
4. Discuss the environmental impacts of Native American game hunting practices.
5. Discuss the impact and influence of the European settlers on the environmental practices of Native Americans.
6. How were Native Americans able to maintain ethical standards without the use of a written language?
7. Discuss the Native American systems of land ownership in light of environmental considerations.
8. What are the environmental positives and negatives of living according to Native American ethics?

Ethical Views on Pollution According to Non-Religious People

Abdulrhman Alsomali

Introduction

Environmental issues such as pollution are not held in high regard by non-religious people. They attach little meaning to environment and how it determines their lives. The ethical positions of those without a given religion on environmental issues are quite different from those who have religious standpoints (Bhat, 2015). Regardless of people's opinions, the issue remains that environment factors crucially in human survival.

This research paper will explore the subject of non-religious people and their views on environment. A number of critical issues related to the subject of ecology and environment will be highlighted to provide guidance on the correct position on ethics and religion.

Nature and Anthropocentrism

An important aspect that sheds light on the views of non-religious people on matters related to the environment is the concept of anthropocentrism. According to anthropocentrism, human beings have more value compared to nature. Humans are regarded as the main components of the ecosystem, which is inclusive of factors such as water bodies, animals, and plants. Natural dynamics agree with this school of thought and show the connection provided by nature to various inhabitants such as animals, human beings, and plants (Chiras, 1992). Introducing pollution in the equation creates a different situation, resulting in poor functioning of natural elements. Environmental conservation is inevitable in creating a sustained system that abides by the principles of natural integration. Pollution is undoubtedly a major threat to the environment and affects both nature and human beings in equal measure.

The ethical positions of those with no religious affinities are not similar to the principles proposed by those with religious backgrounds. Part of this research paper will look into the difference between religious and non-religious ethics by evaluating some of the issues at hand. For example, those with no religious insights may view nature in a way that is very different from others. They may have perspectives or arguments that other people with religious framework of thoughts might not. Depending on which side the issues emerges from, the bottom line is that the environment is a function of life that supports all organisms in the universe. Anthropocentrism is an ethical position taken by non-religious people. Adding spiritual angle in the debate on environment brings out a different picture of perceptions on the topic. Those with a deep spiritual background are better positioned in terms of understanding the environmental dynamics that shape various phenomena in the ecosystem (Elliot, 2011).

Proponents of non-religious views of environment argue that pollution is part of the environment, given that it is designed in the form of a system that eliminates waste materials. A self-sustaining system is therefore associated with the thinking of non-religious people on the issue of pollution. The question of ethical considerations with regards to the environment has taken a back seat within the thinking patterns of those not propelled by religion. Critics claim that such thinking is not accurate and fails to show the suggested evidence. Scientifically speaking, evidence is an important consideration made before quantifying a given research outcome. In the case of pollution, there are no ethical considerations included during the decision making process. Just to mention, the anthropocentrism approach lacks credibility in terms of providing logic and convincing ethical argument.

Scientific logic dictates that nature commands the recycling of waste products and is limited to the idea of continued pollution. An increase in pollution for a sustained period can lead to the emergence of dangerous toxins which can affect the environment (D'Angelo, 2012). There is no way in which nature can receive toxic chemicals for long without breaking down. An example of this is the pollution problems faced by developed nations such as China and America. Both nations have been grappling with the pollution problem after the realization that economic growth in a polluted environment would not be viable. A number of scientists and ecologists have presented evidence demonstrating the importance of environmental conservation.

Proponents of this postulation agree that humans determine the outlook of nature and thus, have an upper hand in the ecosystem. This implies that other non-human components are subject to the manipulation of humans. People with no religious views do not look at the environment as a support system but rather as a less significant part of the ecosystem. Research studies on ecology have shown that nature is an essential element in human life and can lead to disastrous consequences if not properly managed. For example, cutting down trees affects the environment by killing organisms that depend on them for survival. In addition, soil loses fertility as a result of poor composition of organisms. Evidently, lack of proper care of environment leads to problems in terms of human survival.

Environmental experts contend that nature offers the right conditions for human survival. Therefore, nature is a support system that provides humans with provisions such as food, shelter, and water. Life demands that human beings use these provisions for survival purposes. In contrast to this argument is the perspective offered by non-religious people. The views of non-religious people contradict with logical reasoning based on ecology (Grandjean, 2013). Ecological patterns demand a sense of coordination between those within the ecosystem.

A disruption of patterns of ecosystem would mean that human beings will end up suffering. There is no dividing line between nature and human survival. From this reasoning, it is evident that the perspectives of those who are opposed to environmental conservation are skewed and fail to align with reality. The value based argument regarding humans and the ecosystem has provided differing perspectives that lead to the same conclusion. Proposals on anthropocentrism have not been effective in terms of proving the point of their argument. There is no way that nature and human beings can be divided and as seen from the above reasoning, nature is an integral part of the ecosystem.

Instrumental and Intrinsic Value

Non-religious people have different opinions on the value associated with environment. For example, the instrumental value approach looks at nature as a connected system that provides benefits to people and other stakeholders in the ecosystem. Despite this logical reasoning, people without religious views look at the environment as a system that offers value to a selected few while eliminating others in the process. Their approach is meant to show the aggressive side of nature that humans ought to avoid (Gottlieb, 1995). For this reason, non-religious people do not see the importance of ecosystem in terms of supporting life.

Pollution, according to these people is not a threat, as no value is associated with the environment. It is imperative to note that ethical considerations with regards to ecology follow the outlined principles of developing a positive relationship with the environment. The creation story is another aspect that can be used to oppose the proposals offered by people with few or no religious standpoints. During the course of creation, humans were provided with the autonomy to command all nature including animals and water. This account shows that God intended human beings to take care of the environment. On this point, it becomes evident that the thinking patterns of non-religious people are wrong and their arguments fail to live up to the expectations of ecological principles (Neimark & Mott, 2011).

Intrinsic value on the other hand, means the importance of a given natural element and how well it can provide a sustained mechanism.

Ecologists argue that the environment has the ability of providing a self-sustained system that is not possible with other components of nature such as humans and animals. It can be said that the opinions of non-religious people in relation to pollution are misguided and do not represent the ideals of nature. Studies conducted on the impact of pollution on nature shows that the latter has had a negative impact on the environment for a prolonged period. The resolve by governments and companies around the world to comply with the environmental laws of the United Nations is an example that shows the detrimental effects of pollution on nature. Strong ethical framework should be attached to consideration of matters related to the environment for the benefit of human beings.

Therefore, ethical considerations associated with the environment are supposed to mean that humans have an obligation towards developing nature. Nature is part of human and animal life and therefore, should be held in high regard at all times. Opinions of non-religious people have no credible or irrefutable foundation regarding explaining other ways in which human life can be sustained on the universe without the support of the ecosystem.

Pollution, as mentioned in the previous section, is harmful and affects the long-term functioning of the environment (Schaeffer, 1972). As scientists have discovered in recent years, threats to environment such as pollution have damaged the ozone cover that protects the ecosystem from dangerous UV rays. According to estimates presented in various environmental conventions, the world is in a much worse condition than in the past when such rapid industrialization as today had not taken place. It is for this reason that pollution and other threats affect the functioning of the ecosystem.

Moral Aspect of the Environment

Morality issues associated with the environment have often been used by those opposed to the idea of being accountable to the environment.

Pollution as a threat to the ecosystem should not be ignored to avoid further problems. Ecologists have already proven the fact that the environment holds value in relation to the needs of human beings and animals. This implies that the issue of environmental morality is realistic and should not be taken for granted. Taking into account the description of creation mentioned in the previous section, every signal indicates that human beings need to realize the moral issues related to the ecosystem (Stadel & Rhoades, 2008). Understanding these principles is the first step towards enhancing harmony between people and the ecosystem. A case such as pollution causes disharmony between the environment and humans.

In other words, morality helps in promoting a sense of environmental cooperation that is much needed for efficient operation of societies. A detailed review of ethical considerations emerging from environmental issues shows that ecological issues play a vital role in the sustenance of life. Observing the moral principles of the environment is a sure way of showing respect for the environment. Like other issues that can be attached to moral and ethical considerations, the environment is one in the list and presents a range of diverse perspectives.

Those opposed to the concept of moral obligation towards the environment lack an understanding of how ecological issues are connected. The relationship between various components in the environment is what makes them a powerful force even within the sphere of human activity. Thus, logic dictates that the environment is comprised of numerous issues that raise the question of moral integration. Sufficiency is an issue when it comes to discussion on moral tenets of a given environment (Sherlock & Morrey, 2002).

A close review of the ecosystem shows an increased need for environmental conservation that supports sufficiency within the chain. A less sufficient ecosystem may not be favorable to even animals and plants, let alone human beings. Sufficiency, therefore, adds more evidence on the importance of merging moral issues with the environment.

A self-sufficient ecosystem is an ideal location for organisms and human beings to exist healthily. Disrupting the system through pollution and other causes compromises the value of the environment. Destruction and loss of environmental cover hurts the inhabitants in the long run.

Resource availability and sustainability is another factor that comes into play when reviewing the moral considerations placed on environment. Sustainability is the signal that affirms that the environment is functioning well without major threats. Pollution does not bolster the environment but instead hampers the ecosystem from accomplishing various objectives. Global warming and extinction are some of the consequences associated with environmental pollution. Now a time has come when pollution has in fact increased manifold, regardless of provisions set out at climate conventions and commitments to them. It is unfortunate that stakeholders have failed in developing moral principles that are evident in the natural ecosystem (Van & Vorster, 2012).

Biologists argue that morality is subject to the response patterns of human beings towards the environment. The damning statistics on environmental destruction in various parts of the world show the scope of the situation. Finally, equal access to resources is another issue that shows the moral and ethical considerations that emerge from analyzing the environment. The fact that all components of the ecosystem such as plants and animals get the required resources for survival consolidates the viewpoint endorsing moral reasoning of the environment.

Ecology and Ethics

The principle of ecology and ethics is based on appraising the role of humans in environmental conservation initiatives. In the case of non-religious people, their actions show total disregard of environmental principles that foster balance in the ecosystem. Actions such as permitting or actually accelerating pollution and interfering with the system of nature affect the environment—ollut irreversiblydand therefore should not be encouraged. The overview of ethics and ecology is a reminder of the voluminous work still remaining, in terms of protecting the environment (Sherlock & Morrey, 2002).

Those with inadequate religious conviction cannot comprehend the spirituality behind the ecosystem. As a home meant for human beings and animals, the earth means a lot to those residing in it. For example, it would be reasonable for stakeholders to promote environmental conservation rules that address urgent needs. Provisions barring all forms of environmental destruction should be included in the debate revolving around the environment and associated ethics.

The common interests binding nature and humans can be mentioned with regards to ecological ethics. A good way of implementing ethics would be improving the relationship between nature and the elements that depend on it (Walmsley, 2012). Being sensitive to the needs of the environment shows a resolve towards harnessing various subsystems of nature. Eventually, everything will depend on how stakeholders play their roles in terms of shaping the environmental sphere. A strained relationship is bad for both parties and beings negative consequences on board. Systematic thinking that looks into the needs of nature should be used when interacting with it. Doing this not only protects the interests of nature but all the sub-elements that depend on it for survival.

Conclusion

Ethical considerations made with regards to the environment have created mixed reactions among proponents and those opposing environmental ethical issues. As evident from the discussion, ethics should be closely related to the role of humans in protecting the environment. Religious views, or their lack, have also informed perceptions regarding the environment and as pointed in this research paper, have created a new set of issues. The debate on value associated with humans and animals contradicts with the opinions of those with different views on the right environmental structure. Morally speaking, the obligatory perspectives of the environment must be embraced by humans. An assessment of intelligence capabilities shows that humans are more endowed as compared to other elements in the environment such as flora and fauna. Deliberations are inevitably called for, when incorporating the issue of ethics in environmental issues.

References

Bhat, P. R. (2015). *Religious ethics, General ethics, and engineering ethics: A reflection.* In S. Sundar Sethy (Ed.), Contemporary Ethical Issues in Engineering (pp. 99–109). Hershey, PA: Engineering Science Reference. doi:10.4018/978-1-4666-8130-9.ch007.

Chiras, D. D. (1992). *Lessons from nature: Learning to live sustainably on the earth.* Washington, D.C: Island Press.

D'Angelo, J. (2012). *Ethics in science: Ethical misconduct in scientific research.* Boca Raton, FL: Taylor & Francis.

Elliott, K. C. (2011). *Is a little pollution good for you? Incorporating societal values in environmental research.* New York: Oxford University Press.

Grandjean, P. (2013). *Only one chance: How environmental pollution impairs brain development—and how to protect the brains of the next generation.* Oxford: Oxford University Press.

Gottlieb, R. S. (1995). *This sacred earth: Religion, nature, environment.* New York: Routledge.

Neimark, P., & Mott, P. R. (2011). *The environmental debate: A documentary history, with timeline, glossary, and appendices.* Amenia, NY: Grey House Publishing.

Schaeffer, F. A. (1972). *Pollution and the death of man: The Christian view of ecology.* London: Hodder & Stoughton.

Stadel, C., & Rhoades, R. E. (May 01, 2008). Listening to the mountains. *Mountain Research and Development, 28*(2), 183–184.

Sherlock, R., & Morrey, J. D. (2002). *Ethical issues in biotechnology.* Lanham: Rowman & Littlefield Publishers.

Van, W. J. H., & Vorster, N. (January 01, 2012). An introduction to the theological politico-ethical thinking of Koos Vorster. Original research. In *Die Skirling, 46, 1, 1–10.*

Walmsley, G. (April 01, 2012). Authentic faith in a "secular age": McCarthy and Lonergan on the dialectic between sacralisation and secularization. *Missionalia: Southern African Journal of Mission Studies, 40,* 24–63.

Questions on Non-Religious People's View on Environmental Issues

1. What is harmony between humans and the natural environment according to non-religious people?
2. What are common ethical beliefs about environmental issues among non-religious people?
3. What ethical stands are often different between non-religious people, followers of Abrahamic religions, and followers of other religions?
4. How has your ethical views of environmental issues changed since studying these chapters (3 & 4)?
5. What do you want to do differently?

Chapter 5

Sustainability, part a

Sustainability is a word that is bandied about a lot, but what do people mean when they talk about sustainability, or about living sustainably? We like to think that living sustainably means that we are living, and using resources in a way that can be continued indefinitely, but is that really true? We are using natural energy resources, such as oil, natural gas, and coal, to generate power, enabling easy and quick transportation, and making of things that we will use for a while and eventually discard. We know that oil, natural gas, and coal are finite resources that will not regenerate in millions of years. Still, we use them as though they will last forever. Sure, there are efforts to replace some conventional sources of power drawn from exhaustible resources with solar and wind generated energy, but as the world population grows and developing countries demand more of the power-hungry tools of the "civilized" life, the increasing need for power and fuel are outpacing the "green" replacements.

Nevertheless, we must at least make efforts to try to strive for sustainability. Even if we cannot truly attain the goal, we can at least move in that direction, and extend the amount of time for which we have the current natural energy resources available. In this chapter, two authors will tackle sustainable practices in two important areas: land transportation, and manufacturing—areas are responsible for consumption of a large portion of the energy resources that are used around the world.

Sustainable Practices in Land Transportation

Naif Aldawsari

Introduction

This research paper provides description of numerous sustainable practices in land transportation. First of all, ample attention has been paid to the terms 'sustainability' and 'transportation sustainability' for better understanding of the topic being studies. Next, numerous practices have been described, along with examples of their implementation in different countries all over the world: in the USA, Canada, and countries of Asia and Europe. Sustainable practices of inland transportation have several major aims: environmental protection and coverage of social and economic needs of human and goods transportation. The above-mentioned practices have numerous directions like saving of the environment through using high quality fuel and alternative means of transportation; numerous pricing practices; traffic calming initiatives; wide implementation of technological solutions; a more effective and efficient land use; promotion of walking and bicycling; and increase of the people's awareness about the necessity of above-mentioned initiatives.

Sustainability and Its Reflection in Practices of Inland Transportation

In order to understand the nature of sustainable practices in land transportation, this research paper provides some generalized description of sustainability. This term undermines development and implementation into real-life strategies, aiming to reach a balance of environment, society, and social needs and economy. These processes enable providing equitable life, existence, and resources for people, flora, and fauna. The Sustainability Act of Oregon provides the following definition of the above-mentioned term: "using, developing and protecting resources in a manner that enables people to meet current needs while providing for

68

future generations to meet their needs, from the joint perspective of environmental, economic and community objectives" (Centre for Environmental Excellence by AASHTO, 2009, p.G1). It should be mentioned that all parts of sustainability (environment, society, and economy) should be balanced equally. Thus, quality of sustainable procedures is much more important than their quantity, i.e. development stands over growth (Oregon Department of Transportation, 2005).

Transport sustainability undermines creation and realization procedures, enabling increase of environmental and human safety and security. The majority of sustainable practices in land transportation are based on initiatives directed at saving nature and decreasing the amount of harmful emissions like carbon dioxide. These initiatives are implemented by price regulation procedures; precise planning of transportation; improvement of logistics; implementation of IT technologies; searching of alternatives for fuel and fuel-using mechanisms; facilitating bicycling; and increasing public awareness concerning the necessity of realization of sustainable practices in land transportation. Below, all these initiatives will be described in a more detailed way. Additionally, these initiatives have the aim of solving considerable transport problems that arise in the course of time due to considerable social and technological developments. These problems are the following: insufficient land-use policy and lack of coordination of the transportation process; inappropriate infrastructure; insufficient control of harmful emissions; and considerable increase of the amount of automobiles on roads caused by such factors as urbanization, motorization, and economic development (Hayashi, Doi, Yagishita, & Kuwata, 2004).

Hence, sustainable transportation is transportation that, on the one hand, covers basic transportation needs of people safely and efficiently and, on the other hand, avoids harm to the environment and to people's health. Minimization of harmful emissions, lowering of the use of nonrenewable resources, and decrease of land use are also among main priorities of sustainable transportation. At the same time, such transportation should correspond to needs of the economy on timely and cost-saving land transportation for goods and people.

Viewed superficially, reaching all these goals simultaneously may be a rather difficult and complicated task. However, precise understanding of the necessity of implementation of practices of sustainable transportation, accurate development of these practices, and their timely realization will make reaching these goals possible. Accommodation of demand of human mobility without increasing the amount of vehicles or without expanding infrastructure can be reached by development and implementation of ride-sharing and bicycle sharing plans. Increasing effectiveness and efficiency of inland transportation can be obtained by applying incident management and implementing intelligent transportation system with the use of modern technologies. Besides, such initiatives as expansion of streets and highways, as well as development of bicycle and pedestrian infrastructure are of top importance. Other sustainable practices in inland transportation include the use of renewable fuels, which minimizes harm to the environment; work with materials that can be obtained locally, minimizing the need for transportation of these materials; and use of high quality materials for building roads and other objects of transport infrastructure.

Main Goals of Sustainable Practices in Land Transportation

One of the primary goals in realization of sustainable practices in land transportation is saving the environment from the harmful effects of using transport. This goal can be obtained by developing procedures, which will "avoid resource depletion, avoid adverse impacts to the environment and society, and is affordable" (Oregon Department of Transportation, 2005, p.G4). The main problem of transportation is emission of harmful substances into the atmosphere. These harmful substances are carbon dioxide (CO_2), halocarbons (HCFCs), nitrous oxide (N_2O), sulphur dioxide (SO_2), methane (CH_4) heavy metals, etc. Transportation leads to an increase of greenhouse effect on earth. Another problem is environmental depletion, especially oil depletion. It must be mentioned that the USA is the country that produces the largest amount of greenhouse gasses. This is connected with the fact that the population of the United States consumes the largest amount of energy. According to the information obtained from the "Oregon Transportation Plan," "the greenhouse gas emissions from the United States currently account for almost one-quarter of the worldwide total" (Oregon Department of Transportation, 2005).

Procedures enabling control over emissions are implemented all over the world. For example, effective and efficient strategies are developed and realized in Asian countries. Thailand and Malaysia, for example, focus their efforts on numerous restrictive measures, which allow reduction of emission of carbon dioxide and sulphur dioxide (Hayashi et al., 2004). These countries have widely implemented numerous strategies for strengthening inspection of automobiles. Special advanced test stations for analysis of harmful emissions are also used in these countries. All these practices aim to change people's perceptions regarding using low-cost low quality fuel to using high quality but more expensive fuel. For example, the Chinese government facilitates using a water diesel emulsion on personal automobiles and on public service vehicles in Beijing and Shanghai. Emission of nitrogen oxide and carbon oxide from application of water diesel fuels is considerably lower than emission of these substances from common diesel fuel.

Sustainable transportation should be promoted and sponsored by the government. In the article "Sustainable Transportation Gains Momentum at Rio+20," Hughes and Replogle (2012) state the following: "more and better-targeted investment in transport would be beneficial, but these need to be accompanied by governance reforms to facilitate improved operations of transport systems, with more accountability for performance."

Sustainable practices in land transportation are developed and implemented by DOTs (departments of transportation). In "Transportation and Sustainability Best Practices Background", it is stated that new approaches and technical innovations should be widely used for supporting and enhancing a sustainable future. Realization of sustainable practices lies on departments of transportation. These departments should correspond to the following social, economic, and environmental needs: "need to adapt to changing economics, natural resource supply, and carbon constraints (including regulation and potential federal policy related to climate change), and the need to mitigate these changes as part of a broader societal effort to address issues such as climate change, water availability and quality, security concerns and rising cost of energy" (Centre for Environmental Excellence by AASHTO, 2009). Topmost priority is given to solving issues of environmental pollution, greenhouse effects, and climate changes. There are two main directions of solutions of these problems: adaptation to current environmental changes, and their mitigation. Adaptation is reflected in the use of high quality materials for building roads and other objects of infrastructure, to maximize their ability to withstand natural disasters (hurricanes, heavy rains, sandstorms, etc.). Mitigation is reflected in searching for and implementation of technological means and methodologies of decreasing emission of harmful gases and substances. The example of such mitigation initiatives is a requirement to use low carbon materials for construction of objects of infrastructure. In any case, departments of transportation should see with a clear vision that any current activities will lead to a considerable improvement or deterioration of life of future generations.

Pricing Practices in Land Transportation

The main goal of development and implementation of pricing practices in land transportation is a considerable decrease of congestion and use of automobiles, especially in big cities. These practices comprise the following initiatives: pricing of roads for peak congestion period, toll roads, alternative fuel revenue tax, etc.

The so-called congestion pricing is an initiative that assumes taking money from owners of vehicles for using certain particular roads during the peak congestion period. The department of transportation usually uses this method only on certain mostly congested roads.

The second initiative that can be implemented as one of sustainable practices of inland transportation is alternative fuel revenue tax. This initiative considers levying payments on vehicle distance travelled: "distance-based car insurance and distance-based car registration fees convert insurance and registration fees to a variable cost related to annual miles driven" (Oregon Department of Transportation, 2005, p.G15). This will lead to a considerable lowering of the use of automobiles.

The next pricing initiative is increasing prices on fuel, also known as carbon taxes. This practice can change people's perceptions of fuel, motivating them to use alternative means of energy. According to official researches performed by Elizabeth Deakin and published in the *Oregon Transportation Plan*, "increasing fuel prices by a rate of 3 per cent per year would result in a 20 percent reduction in global warming by 2020 and 35 per cent reduction by 2040" (Oregon Department of Transportation, 2005, p.G11).

The initiative that stipulates buying fuel-efficient automobiles represents one more sustainable practice of inland transportation. In some American states, authorities provide more than US$1,000 in encouragement and credits. Some authorities have given a priority to public transport instead of personal automobiles. As per "Sustainable Transportation Practices in Europe," such positions "encourage people to use cars thoughtfully and sparingly" (U.S. Department of Transportation, 2011, p.28). Some people decide to buy new hybrid cars while others decide to use environmentally-friendly fuel. Departments of transportation develop and implement numerous programs all over the USA. According to their initiatives, "urban buses, light rail, intercity rail, and intercity bus are understood to have generally lower GHG emissions per passenger mile than conventional options and are therefore considered important emissions reduction strategies" (Centre for Environmental Excellence by AASHTO, 2009). In his work *Sustainable Transport Best Practices and Geography: Making Connections*, Berry Wellar proposes an initiative that considers streets in big cities having special sections only for buses, and lanes for heavy traffic times and peak hours. At the same time, "buses in cities should have legal and acknowledged right-of-way for turns at intersections and into traffic, regardless of street signals and markings" (Wellar, 2007).

Departments of transportation implement pricing policies all over the country. These policies consist of collecting information concerning current prices on the fuel marker, collecting data concerning congestion rates in different cities and areas, and receiving data relating to the amount of vehicles in different locations and the amount of harmful substances emitted into the atmosphere. This information is analyzed by specialists who develop strategies for implementation of effective and efficient pricing policies all over the country. Much attention is paid to policies, which in turn stipulate increasing quality of used fuel. This leads to growth of fuel efficiency, decreasing the amount of used fuel for transportation and lowering the negative impact of harmful emissions on the atmosphere. One of the most notable sustainable practices in land transportation was implemented in Washington, DC. Special Washington's Traffic Choice Study was performed. According to these investigations, the "role that time-of-day variable road tolling can play in bringing more balance to transportation supply and demand" is remarkable (Centre for Environmental Excellence by AASHTO, 2009). Special Global Positioning System (GPS) tolling meters were put on vehicles in Washington, DC. Expenses on deduction tolls were covered from the local budget. In this research, variable tolls were changed for the use of drivers on major roads of the city. Researchers collected information about driving patterns of participants before starting the experiment and after its implementation. This experiment shows considerable changes in transportation after implementation of the above-mentioned initiative. The amount of total traveled distance decreased by more than 10 percent, while the amount of automobile trips decreased by about 7 percent (Centre for Environmental Excellence by AASHTO, 2009). The result of Special Washington's Traffic Choice Study was successful implementation of so-called high-occupancy toll lanes all over the state.

One more pricing sustainable practice in land transportation is mileage-based insurance of automobiles. This initiative has the aim of lowering distances of transportation and the number of vehicle miles traveled. This practice has successfully worked in territories of the USA for more than 10 years. According to Centre for Environmental Excellence by AASHTO (2009), "in 2002, the Texas Department of Insurance Commissioner approved rules to enable insurers to offer automobile insurance plans that allow consumers to purchase insurance coverage on a per mile basis." Two different kinds of mileage-based insurance were presented to customers: a cent per mile rate and an unlimited mileage rate. It should be mentioned that a car becomes uninsured when it is driven more than the amount of purchased miles. At the same time, people receive premium for any unused mileage.

Car sharing programs aim to decrease the number of automobiles all over the country. Nowadays, this program is working successfully in Portland, Los Angeles, and San Francisco. People are not obliged to buy a car for occasional trips. At the same time, company-owners bear expenses on maintenance and insurance of their automobiles. The report of U.S. Department of Transportation titled "Sustainable Transportation Practices in Europe" describes the main goal of these programs: "car-sharing programs designed to provide households the convenience of occasional automobile use without necessitating ownership or costly rentals" (U.S. Department of Transportation, 2011, p.22).

Traffic Calming

Numerous programs have been created to make traffic calmer. The most outstanding ones are reflected in enforcements against driving when drunk and improved designs of highways. These initiatives make roads safer and more sustainable. High quality of highways and using of high quality materials also lead to a considerable improvement of urban livability (U.S. Department of Transportation, 2011). It is notable that some authorities prefer to upgrade existing roads (by making them wider and safer) instead of building new ones. This saves both money and land.

Besides, among other traffic calming initiatives can be named increase of parking places. Parking in big cities reduces road traffic incidents. The practice of traffic calming is widely implemented in shopping areas of big cities.

Use of Technological Solutions as One of Sustainable Practices in Land Transportation

These days, technological solutions are widely used for increasing effectiveness and efficiency of the process of transportation. Some of these solutions help people to improve automobile movement by providing information concerning locations, weather, facilities, and schedules. For example in Oregon, global position systems are introduced on passenger wagons for providing people with the latest information about their current position (Oregon Department of Transportation, 2005). People also have the opportunity of receiving information about schedules, cost of transportation, etc. This appliance helps people choose different means of transportation and provides increasing popularity to common modes like buses and trains.

Moreover, technological innovations, which ensure saving of fuel and protection of the environment, are widely implemented into automobiles and railroads. For example, a special railroad car that works on acid batteries has been developed and made by the Green Goat Company. This hybrid emits up to 90 percent less nitrous oxide than common railroad cars (Oregon Department of Transportation, 2005).

Numerous intelligent transportation systems (also known as ITS) are widely used for helping drivers to find the shortest way from the point of departure to the point of destination. These technologies also decrease driving on congested roads during high traffic rime periods.

Sustainable Practices in Land Use

Departments of transportation perform extensive research or change infrastructure all over the country for reduction of the necessity of transportation of people. They try to remake cities and towns and to increase the amount of people who are not obliged to drive long distances from their living places to their working places, shops, hospitals, schools, or entertainments (theatres, cinemas, etc.). According to the work "Transportation and Sustainability Best Practices Background," there is a direct connection between high intensity of land use and transit of people from one point to another. Departments of transportation are developing different initiatives for implementation of a more effective and efficient planning. As an example of effective and efficient implementation of sustainable practices in land use, one can take an example of work of the Department of Transportation in Florida, California, and Virginia. Governments of these states have developed and realized initiatives for making roads larger, improving highways, and using new materials and technological solutions enabling transportation facilities to withstand heavy weather conditions. These initiatives aim to save people's life in extremal situations. The Department of Transportation in Virginia successfully performs identification and uses the score system for determining the condition of bridges. Moreover, this department facilitates building and properly maintaining hurricane evacuation routes. These initiatives are meant for preventing destruction; for performing detection of weak chains in transportation infrastructure in due time and rectifying problems that may arise; and in providing a timely, effective, and efficient response to any extreme conditions.

It should be mentioned that local departments of transportation closely cooperate with local agencies for developing more community-oriented land use strategies and making more effective and efficient investments into local infrastructure. These initiatives concern both urban transit systems and commercial lands. Below, examples of cooperation of the department of transportation with local agencies for implementation of sustainable practices of inland transportation are provided.

The Department of Transportation of Minnesota, the Federal Highway Administration, and the Metropolitan Council have joined their forces for the lanes project. This project has the goal to obtain optimal performance of high occupancy vehicle lanes. There is considerable traffic congestion on these lanes during high traffic hours. Thus, the above-mentioned agencies have decided to transform lanes into tolled ones. High traffic tolls are deemed to optimize traffic at least for the following 30 years.

Interchange Area Management Plan is one more long-term initiative of the department of transportation. It aims to improve highway interchanges in a long-term perspective (at least for the next 20 years). Supporting of the network of local streets and protection of highway interchanges are of top importance in this plan.

Sustainable practice in land use also has the aim of protecting environment from the harmful influence of various emissions. Different construction solutions are used for this. An example of such solution building is a terrace on slope for reduction of land disturbance. Transportation between Canadian cities of Vancouver and Whistler was organized in such manner in 2010. This project is named the Sea-to-Sky Highway Improvement Project: "by constructing a retaining wall in between the two tiers of traffic, workers were able to narrow the construction right of way. Sea-to-Sky extensively used Mechanically Stabilized Earth" (Centre for Environmental Excellence by AASHTO, 2009).

Bicycling and Walking

Replacement of automobiles by bicycles and of driving by walking will certainly lead to considerable ecological improvements. Walking and biking do not require any fuel. Only the human body and a bicycle are needed. It should be mentioned that, as per researches provided in the *Oregon Transportation Plan,* nothing less than 98 percent of energy used by the automobile is used to move the car and only remaining 2 percent are used to move people. That means that the greatest part of fuel is consumed not by effective work. Moreover, by walking and bicycling individuals perform physical activities and consequently improve their health.

City authorities should implement ample procedures for improvement of walking and bicycling opportunities. More sidewalks and bike lines should be made. Streets should be safe and easy to cross for bike riders and pedestrians. Decreasing of the amount of highways and lays will make walking easier, safer, and more convenient.

Increasing of People's Awareness Concerning Sustainable Practices in Land Transportation

The above-mentioned practices in land transportation should be widely supported by increasing people's awareness of them. It is extremely important to describe to people the necessity of implementing new practices and explain that these practices can lead to a considerable improvement of health and overall well-being, reduce emission of harmful substances into the atmosphere, decrease greenhouse effect, and save nonrenewable sources. The awareness will facilitate implementation of sustainable practices in land transportation and increase their effectiveness and efficiency. Examples of such awareness plans are the ME3 (the USA) and the Travel Smart (Australia), and Travel Wise (the UK). For example, the Travel Wise program is widely implemented all over the United Kingdom and aims "to educate the public to think before traveling, to walk or bike when possible, to use transit effectively, and to plan auto travel to chain trips, avoid congestion" (U.S. Department of Transportation, 2011, p.29).

Conclusion

This research paper provides a brief overview of sustainable practices in land transportation and their necessity for protection of the environment and coverage of social and economic needs of inland transportation. Departments of transportation closely cooperate with numerous agencies all over the country for development and realization of complicated transportation projects. The goals of these projects is to increase quality of used fuel for lowering a negative impact on the environment, to use alternative means of transportation, increase effectiveness and efficiency of land use, promote implementation of numerous information technology aided applications in the transportations sphere, increase people's awareness concerning the necessity of the above-mentioned initiatives, and encourage bicycling and walking.

References

Centre for Environmental Excellence by AASHTO. (2009). *Transportation and sustainability: Best practices background.* Prepared for Transportation and Sustainability Peer Exchange, May 27–29, 2009. Gallaudet University Kellog Centre: Washington D.C. Retrieved from http://environment.transportation.org/pdf/sustainability_peer_exchange/AASHTO_SustPeerExh_BriefingPaper.pdf

Friedl, B., & Steininger, K. W. (2002). Environmentally sustainable transport: Definition and long-term economic impacts for Austria. *Empirica*, *29*(2), 163–180.

Hayashi, Y., Doi, K., Yagishita, M., & Kuwata, M. (2004). Urban transport sustainability: Asian trends, problems and policy practices. *European Journal of Transport and Infrastructure Research, 4*(1), 27–45.

Hughes, C., & Replogle, M. (2012, May 9). Sustainable transportation gains momentum at Rio+20. *Opednews.com.* Retrieved from http://www.opednews.com/articles/Sustainable-Transportation-by-sustainable-prospe-120904-936.html

Johnson, C. (2014, November 19). How to start a bike kitchen. *Resilience. org.* Retrieved from http://www.resilience.org/stories/2014-11-19/how-to-start-a-bike-kitchen

Kassens, E. (2009). Planning for sustainable transportation: An international perspective. *MIT Journal of Planning.* Retrieved from http://web.mit.edu/dusp/dusp_extension_unsec/projections/issue_9/issue_9_kassens.pdf

Larusdottir, E. B., & Ulfarsson, G. F. (2013, September 9). Effect of driving behavior and vehicle characteristics on energy consumption of road vehicles running on alternative energy sources. *International Journal of Sustainable Transportation, 9*(8), 592–601.

Northern Territory Government. (2009). *Roads and the environment: A strategy for sustainable development, use and maintenance of Northern Territory roads.* Retrieved from http://www.transport.nt.gov .au/__data/assets/pdf_file/0009/21042/roads_environment.pdf

Oregon Department of Transportation. (2005). *Oregon transportation plan update.* Retrieved from http://www.oregon.gov/ODOT/TD/TP/ docs/otp/sustaintransdev.pdf

Oregon Department of Transportation. (2008). *ODOT sustainability plan.* Retrieved from http://www.oregon.gov/ODOT/SUS/pages/ sustainability_plans.aspx

Roberts, D. (2010, May 6). Does 'sustainable transportation' mean better cars or fewer cars? *Grist.org.* Retrieved from http://grist.org/ article/2010-05-05-does-sustainable-transportation-mean-better-cars-or-fewer-cars/

U.S. Department of Transportation. (2011, November). *Sustainable transportation practices in Europe.* Retrieved from http://international .fhwa.dot.gov/Pdfs/converted_to_html/sustainabletransportation/ SustainableTransportation.cfm

Wellar, B. (2007*). Sustainable transport best practices and geography: Making connections.* San Francisco, CA: University of Washington.

Walljasper, J. (2014, October 14). Bring home best-in-the-world ideas to make sure your city thrives. *Resilience.org.* Retrieved from http://www.resilience.org/stories/2014-10-14/bring-home-best-in -the-world-ideas-to-make-sure-your-city-thrives

Yang, F., Jin, P. J., Cheng, Y., Zhang, J., & Ran, B. (2013, July 15). Origin-destination estimation for non-commuting trips using location-based social networking data. *International Journal of Sustainable Transportation, 9*(8), 551–564.

Sustainable Practices in Manufacturing

Hammad Alonazi

Over the last few years, rapid industrialization has resulted in increased environmental issues like harmful and irreversible changes in climate and scarcity of various natural resources. Industries have awakened to these looming threats and have started showing increased awareness of the need to employ sustainable practices to ensure sustainable manufacturing. However, the present efforts are insufficient to overcome these grave environmental issues. Various international organizations like the Organization for Economic Co-operation and Development (OECD) have been encouraging initiatives for reducing the emission of greenhouse gases to control global warming. Thus, employment of sustainable practices has become an issue of major concern for human beings across the globe in this current age of rapid industrialization and depleting natural resources. Sustainable development, a key factor for the progress of future generations, can be achieved only if industries across the globe adopt such practices promptly and consistently. Sustainable development, achieved by implementing sustainable practices, aims to address the economic, environmental and societal issues in a holistic and concurrent manner along with achieving the desired development objectives.

Manufacturing units are now increasingly apprehensive regarding sustainability. Many segments like manufacturing, design, and engineering have already implemented sustainable practices to some extent. The impact of manufacturing processes on nature is increasingly being recognized and addressed. Implementing sustainable practices is, no doubt, a tough and complicated task and includes various aspects like engineering, technology, financial concerns, safeguarding environmental concerns, societal aspirations, governmental schemes, rules and regulations and well-being and safety of the surrounding society and people. Manufacturing segment, particularly, needs to balance and

82

integrate the aims related to environment and profitability. The manufacturers must keep themselves constantly updated with pertinent, reliable, and consequential knowledge related to sustainable manufacturing practices that are being discovered and implemented by firms across the globe if there has to be an improvement in sustainability in the manufacturing sector and overall sustainable development is to be achieved (Rosen & Kishawy, 2012).

Having discussed the importance of sustainable practices in manufacturing segment, this paper will discuss the background, need, and advantages of sustainable practices; influencing factors; different sustainable practices that can be adopted by firms; how they are measured; firms that promote sustainable practices; and how various well-known companies have benefitted by implementing sustainable practices without affecting their bottom line. The primary aim of this paper is to enhance the reader's understanding and to encourage rapid adoption of such practices in manufacturing industries in the near future.

Background and Definition

Conventional manufacturing practices did not concentrate on sustainability. The concept of sustainable manufacturing practices emerged from sustainable development, which was created during the 1980s to deal with issues related to economic progress, inequalities, effects on environment, globalization, and various other aspects. Sustainable manufacturing was first discussed at the 1992 UNCED conference (United Nations Conference on Environment and Development) held at Rio de Janeiro to encourage governments and organizations to adopt measures for sustainable development. There are many definitions of sustainable manufacturing. The U.S. Department of Commerce defines sustainable manufacturing as the usage of such procedures that reduce the negative impact on the environment, preserve natural resources and energy, ensure safety of society, consumers and employees, and are financially viable. According to Lowell Center for Sustainable Production, sustainable manufacturing can be defined as the production of services and products by means of methods and techniques, which conserve natural resources and energy, are financially feasible, non-polluting, and safeguard the communities, clients, and employees as well as are generally and innovatively worthwhile for all (Rosen & Kishawy, 2012). Overall, sustainable development can be defined as meeting the requirements of the present generation, but not at the cost of forcing the next generations to compromise in order to fulfill their requirements (World Commission on Environment and Development, 1987). Thus, sustainability has been defined and understood in various manners, based on the needs of the different applications.

The Need for Sustainable Practices

Manufacturing industries comprise a major portion that use various resources and generate wastes. The consumption of energy by production units from the year 1971 to 2004 across the globe has increased by almost 61% and constitutes almost one-third of the total use of energy worldwide. Similarly, almost 36% of the total emissions of CO_2 (carbon dioxide) are produced by manufacturing units (IEA, 2007). Being the largest consumers, they also have the capability to emerge as one of the strongest forces for creating a sustainable world. They can design and execute sustainable practices and create goods that can lead to enhanced environment-friendly accomplishments, which calls for a change in attitude and comprehension of industrial manufacturing and espousal of a comprehensive approach to running economic activities (Maxwell, Sheate & van der Volst, 2006).

The changes occurring in the climate is more and more perceived to be due to anthropogenic actions and probably has extremely grave outcomes, leading to resources like water, minerals, and energy now being categorized as scarce resources. The widespread usage of non-renewable resources will possibly affect operations in the future if sustainable practices are not implemented at the appropriate time. The worldwide financial turmoil that surfaced in the last few years has also caused humans to rethink and introspect about the feasibility and eventually the sustainability of the current business procedures that are targeted towards financial growth, but does not address concerns related to the negative effects outside the company or on the natural environment. This aspect has also increased pressures from various stakeholders like investors, clients, society, regulatory associations, governments, suppliers, competitors, and employees for adopting sustainable practices by manufacturing units (Rosen and Kishawy, 2012).

The connection between production operations and the ecology is increasingly becoming evident and accepted. Today, development, efficiency, environmental stewardship and economic prosperity are all identified as valid concerns of the manufacturing companies (Sarkis, 2001). Maintaining a balance between these four aspects have become the strategic objectives of most firms today. The current manufacturing policies usually are framed to take care of procedures, technologies, and merchandise and other aspects like sustainable practices to address the managerial and philosophical facets of production plans. Sustainable practices integrates production with environmental concerns with the objective of achieving overall well-being of the natural environment in the long-term, by employing practices like optimum usage of natural resources and efficient waste disposal systems. This will help to stabilize the present taxing relation between human race and the natural world (Hawken, 2007).

The Advantages of Employing Sustainable Practices in Manufacturing Sector

There are several important advantages resulting from adoption and implementation of sustainable practices by manufacturing firms. Employing sustainable practices in manufacturing sector can help to profitably manufacture goods and at the same time ensure minimum damage to the environment, optimum usage of natural resources and efficient consumption of energy. Various studies have proved that firms making use of manufacturing practices that are efficient and environment friendly stand to gain substantial benefits in comparison to their competitors

(Broom, 2015). Thus, sustainable practices aim to take care of environmental, economical, and societal aspects.

Factors Influencing the Implementation of Sustainable Practices

There are various factors influencing the successful execution of sustainable practices in manufacturing firms, as discussed below (Rosen and Kishawy, 2012).

Information

Wide range of qualitative and quantitative knowledge like the amount and kind of metals used during particular procedures and the amount and kinds of noxious waste produced is essential to make the required implementation and evaluation of sustainable practices. Sometimes it may happen that such required information may not be easily available and it may become difficult to implement sustainable practices, which are dependent on this information.

Organizational structure and values

Sustainable practices like steps in the direction of environmental stewardship are usually handled and managed by specialized personnel and departments without the participation of the top management, which can result in contradictions and conflicting interests. This may dissuade the culture promoting advancement of sustainable practices within the organization.

Methods

Personnel related with decision making and actual execution of sustainable practices are usually not presented with the details about the methods and processes required for the effective, efficient, robust, and consistent application of the company's goals and plans related to sustainable practices. One of the probable reasons for this issue is that the quantity of variables required to be considered for decision making is usually very huge. If the sustainability aims are to be accomplished, the working staff ought to effectively consider the sustainability challenges during decision making and implementations.

Sustainable Practices Employed in Manufacturing Sector

There is a false belief commonly held by manufacturing firms that the actual meaning of sustainability and sustainable practices is to direct efforts towards energy conservation. However, to ensure sustainability, it is essential to incorporate it within all stages of the product's life span. Various tools to aid such inclusions have already been built and employed.

Sustainable Practices during Product Design and Life Cycle Evaluation

The traditional role of a product design within production companies is to confirm that the product fulfills all its required goals depending on customer specifications by incorporating ergonomics, effectiveness, visual appeal, and performance. However, due to the increased attention on employing sustainable practices in manufacturing, the designer is additionally responsible for thinking about the effect of design on environment. Such designs require considering the effect on environment during the entire design phase. This is an important part of implementing sustainable practices in manufacturing. The selection of materials and procedures during design phase is equally significant. Design for environment (DfE) is such a design procedure for products, which takes into consideration the probable impact on environment that can be caused by the product during the course of its life cycle and endeavors to reduce this negative effect by dealing with such problematic issues during the product's design phase (Glavic & Lukman, 2007). This is also termed as eco-design.

The life cycle of a product essentially comprises of five phases— namely, pre-production, production, product delivery, usage, and

discarding or recycling. In order to have an overall and complete comprehension of the effects on environment, the complete life cycle of a
manufactured product or procedure should be contemplated by means
of its life cycle assessment (LCA). LCA enables inclusion of sustainability within the designing process by evaluating all facets related to
a product, like the raw items used and the litter produced by it during
its entire life. LCA has emerged as one of the most valuable tools for
enhancing the functioning of various procedures and techniques with
minimum impact on environment as it analyzes the impact of an item
or a service during all stages to decrease the damage to nature. This
is done by improving and focusing on conservation of resources and
increasing efficiency. LCA basically involves four steps namely defining the objective and extent, analyzing the life cycle stock, assessing
its impact and interpreting the findings (Rosen, 2002). LCA conducted
for currently used processes and design options involves analyzing the
employment of energy and various other natural resources and emission
of waste material into environment and energy wastage (Harms et al.,
2008). LCA enables strategies to be designed for the appropriate design
and selection of manufactured goods, raw materials, procedures, reprocessing, and final discarding of wastes. It can help prevent unnecessary
environmental pollution and aid in endeavors for green design. LCA
has been included within the ISO series 14040 standards (International
Standards Organization, 2006). It is commonly used by companies in
combination with assessment of toxic effects and hazard prospects to
encourage usage of sustainable practices in manufacturing units (Rosen
and Kishawy, 2012). Sometimes, executing a full LCA could be time
consuming and wearisome as it is not necessary that all information
related to the product is accessible during the design phase. To have
a rough approximation of the effect on environment, a simple LCA
has been established depending on the concept of group technology
(Kaebernick, Kara, & Sun, 2003). In this, the various products are clus-
tered together depending on their effect on environment during the various phases of the life cycle. It is also necessary to acquire the end-of-life
(EOL) cost approximations for each of the various alternatives during

the recycling, reusing, or discarding phases. This enables the firm to decide the best alternative for the product. DfE and LCA integrate concerns related to both financial and environmental aspects within the design phase of the product.

Sustainable practice during designing includes the integration of sustainability aims. Involving sustainable practices during design stage is still in primary step. Triple bottom line methodology is one such approach to develop sustainable designs, wherein organizations aim to maintain balance between the conventional financial aims and the society's apprehension for the environment (McDonough & Braungart, 2002). Karlsson and Luttropp's EcoDesign technique (2006) aims at preserving or enhancing the financial outcomes while reducing the effect on ecology. Borea and Wang (2007) defined the relation between employment of quality functions, analysis of life cycle and evaluation of contingents and all these aspects are evaluated with respect to customer's keenness to spend for environment friendly products. Grote et al. (2007) discussed an approach for product development with the help of DFX (design for X) tools like analysis of life cycle and TRIZ (Theory of Inventive Problem Solving) enabling the designing engineer to employ eco-friendly ethics without compromising much on the financial front.

Sustainable Practices for Energy and Resource Consumption

Various sustainable practices have been developed for fields associated with manufacturing like usage of resource and energy. For resource usage, Smith and Rees (1998) explained sustainable practices as a model of using resources such that human requirements are met and the environment is also preserved at the same time enabling the requirements to be met by the present and coming generations. Sustainable practices for energy usage entails the usage and control of energy sources in such a sustainable style that there is enough energy for meeting the fundamental requirements of the present and future generations through resources that are reasonably priced and do not harm the environment in any manner and is well accepted by the people and society.

Sustainable Practices during Production

Sustainable practices employed during production aims to produce goods using resources in a controlled manner and causing minimum pollution. Womack and Jones's (1996) lean principles portray the significance of enhancing the production techniques on a continuous basis by concentrating on removing waste generation. This will result in many benefits like improvement in quality, reduction in wastes, and increased market share apart from being eco-friendly. Manufacturing companies that adopt a holistic approach to oversee operations using lean principles, agility idea, and sustainable manufacturing practices remain financially feasible and succeed in entering new markets. Consumption of energy during manufacturing can be decreased by automatically switching off non-required machines. To maximize the usage ratio of product to resource, reconfigurable and supple products that can be utilized in various applications should be encouraged. Firms are constantly searching for methods to integrate sustainable practices with their industrial processes like sustainable administration, sustenance of environment and triple bottom line manner. The triple bottom line method has been explained by Elkington (1998) as taking care of sustainability by adopting holistic methods based on the ideologies of financial progress, corporate social responsibility, and environmental stewardship.

One of the other sustainable practices that can be used in manufacturing is lengthening a product's life span either by reusing, reproduction, or recycling. Remanufacturing comprises of considerably modernizing or revamping of mechanical tools, machines, or other things to make them almost new or in a reusable form (Glavic & Lukman, 2007). This is much better as compared to recycling, as it involves reusing of parts. However, this may lead to quality and consistency issues (Anityasari & Kaebernick, 2008).

Measurement of Sustainable Practices

Sustainability indicators are required to indicate how a manufacturing firm can influence sustainable development by adopting sustainable practices. These indicators or metrics are required to evaluate how much sustainability has been achieved by the implementation of such practices. Many manufacturing companies have attempted to incorporate sustainable practices within the main decision-making systems. Defining sustainability metrics is, no doubt, a tough challenge. Parris and Kates (2005) have defined and recognized almost 255 indicators identifying sustainability. These metrics range from geographic coverage, capability to be controlled by the key decision makers and the actions and expenses needed to implement these sustainable practices. Stokes (2009) suggested monetizing sustainability depending on including the triple bottom line technique within the manufacturing practice. Aspects like complying to environmental regulations and efficiency of processes allow for outcomes to be measured aided by conventional business goals, but in order to assess the outcomes, they need to be monetized depending on the expected results, prioritization of outcomes, and business functioning. These performance indicators are crucial for enhancing the implementation of sustainable practices in manufacturing units as these steps need well-defined metrics to be assessed, improved, and implemented in future (Hutchins, Gierke & Sutherland, 2010).

Various performance measures have been devised for measuring the success of sustainable practices. Organizations and firms at various international, national, and regional levels have suggested indicator groups or principles. Most of these guidelines are relevant—either partially or completely—for gauging the effectiveness of the sustainable practices used during designing and manufacturing stages for various procedures and manufactured goods. Many organizations have developed and implemented their own pertinent indicator sets for sustainability. GM (General Motors) has developed metrics for assessing the impact of sustainable practices depending on an evaluation of high-tech and latest metrics available for sustainable manufacturing practices. These guidelines recommend implementing almost 30 metrics for sustainable manufacturing practices in six different key fields like consumption of energy, health of employees, impact on environment, waste disposal system, manufacturing expenses, and safety precautions (Feng, Joung & Li, 2010). Ford has developed a sustainability catalog, which includes eight parameters including environment, society, and economics. These parameters have been developed based on analysis of life cycle expenses, assessment of life cycle, sustainable substances, well-being, noise pollution and ability related to portability and various other aspects (Dreher et al., 2009). Walmart has developed a sustainability product catalog based on 15 questions asked to suppliers, meant to motivate suppliers to implement sustainable practices at their end and to assist clients in making informed buying decisions (Ford, 2007). The sustainability directory by Dow Jones evaluates the economical and sustainable performance of the leading 10% of members of the Dow Jones Global Stock Market Catalog with the help of 12 conditions related to environmental, societal, and financial factors depending on the information from the organizations, media reports, and stakeholders (Walmart, 2015).

Global institutions have also developed various useful metrics to measure the impact of sustainable practices. The indicator set of United Nations Commission on Sustainable Development (2007) is made up of mainly 50 metrics, which have been divided into 14 various segments and also consists of principles for application and adaptation of the metrics for developing metrics at national level. The OECD has created a basic collection of indicators for environment that comprises of almost fifty indicators related to a broad variety of aspects related to economic and environment and initiatives or reactions by domestic segments, industries and governments. To assist in its initiatives and to encourage sustainable practices in manufacturing segment, the OECD has also instituted a resource pool that offers technical assistance for SMEs for assessing and construing eighteen basic indicators related to assessing performance of sustainable practices (OECD, 2000). The European Union has also developed a group of indicators related to environment that comprises of sixty markers describing the impact created by man-made activities on the ecology. It is further divided into ten strategic segments like air pollution, changes in climate, efficient disposal of toxic materials, and biodiversity, thus covering majority of the important anthropogenic actions harming the natural environment (EuroStat, 1999).

Organizations Promoting Sustainable Practices in Manufacturing Sector

The manufacturing and services segment of International Trade Association (ITA) aims to increase the reach and knowledge about processes that are environment friendly and resource efficient and can lead to economical and cost efficient goods with the help of SMART (Sustainable Manufacturing American Regional Tour). It regularly holds tours of various manufacturing units that have already employed sustainable practices like wind power and pioneered methods for heat and power administration, and sponsored environment-friendly designs and recycling systems. Thus, SMART enables other business organizations to learn from those who have already adopted these sustainable practices and are running them successfully (Broom, 2015).

The OECD initiated a Project on Sustainable Manufacturing and Eco-innovation in the year 2008 with the objective to accelerate implementation of sustainable practices in industrial manufacturing by circulating present information and aiding the process of deciding benchmarks for various goods and manufacturing procedures. It also encourages the idea of eco-friendly innovations and development of modern, technical answers to environmental issues across the world for short-term to long-term periods (OECD, 2009).

Companies that have Benefitted from the Implementation of Sustainable Practices

Adoption of sustainable practices has led to innovations, which are technologically advanced and also preserve the environment (Broom, 2015).

- Eastman Kodak has decreased its overall energy consumption by 36% since 2002.

- Xerox had set a target to decrease emission of greenhouse gases by 10% by the year 2012. Information on the success of that goal is not currently available.
- Harbec Plastics has started using electric presses, which has reduced its energy expenses by almost 50% and increased the processing rate and cycle duration.
- Toyota and Honda have adopted a lean product development system for supervising their product development. This method allows examination of various design options during the entire product development phase and enables the expenses and advantages associated with sustainable designing to be assessed, which ensures that errors are avoided and quality is improved (Morgan & Liker, 2006).
- Intel has given importance to sustainable practices in its manufacturing procedures and products ahead of commercialization (Harland, Reichelt & Yao, 2008). It has adopted a two-year archetype for developing new products and keeps on interchanging from silicon production technology in one year to microprocessor technology in the second year. This pattern enables introduction of a modern manufacturing procedure within the initial year, wherein experimentation is done like reducing the size of semiconductor and consequently producing more number of semiconductors from a specific wafer or placing more number of transistors in equal gaps. During the second year, there is introduction of a new design for chip or a design that uses exactly the same production method. Every stage offers chance to determine aims and policies to decrease the damage to the environment. Intel works in coordination with equipment and material dealers to enhance the adoption of environment friendly steps within various technologies.

The Path Ahead . . .

Manufacturing systems and practices in future will be probably established on policies that can fulfill instant facility requirements in such a manner that it improves the quality of environment for coming generations and the future business projections of the organizations. The manufacturing units now make use of advanced energy systems to enhance the working cost structures like load cutting and shedding, supervising energy usage and controlling usage of generators, thermal units and HVAC systems. There are various possibilities for enhancing the usage of sustainable practices in manufacturing. Firms should continuously increase their production efficiency instead of operating on only conventional economics. Instead of merely fulfilling legal rules they should plan to transition to higher voluntary actions.

There should be inclusion of much more sophisticated restraints and designing factors related to environment and sustainable practices in order to generate a much wider range of design options, which will also allow assessment of impact of sustainable practices on the intricacies of the project, cost of the product and designing of the process in a much more complete manner based on robust data. The present engineering design tools or its amended version like design related to assembly and manufacturing, Six Sigma Design, matrix used for design configuration and implementation of quality function have not yet fully explored and imbibed the concept of sustainability and effective usage of sustainable practices (Johnson & Srivastava, 2008).

Adopting sustainable practices not only helps companies to boost their image but also brings in new customers and improves the bottom line. The top priority for international organizations and manufacturing companies should be having a united vision, approach, and policy for creating more and better economic prospects along with decreased harmful effects on the environment. The immediate steps implemented on an urgent basis today will encourage more innovation in environment friendly infrastructure and technologies that will help to solve community and ecological issues in the long term, thereby helping to achieve sustainable development. Companies should strongly believe that sustainable practices will help them to improve the quality of their products, increase their market share and ultimately their profits. The standards described by various international organizations should not be considered an end in themselves.

Companies need to reinvent on a continuous basis to be sustainable. Thus, an integrated approach wherein all facets of manufacturing operation are considered is essential to gain maximum advantage from sustainable practices.

References

Anityasari, M. & Kaebernick, H. (2008). A concept of reliability evaluation for reuse and remanufacturing. *International Journal of Sustainable Manufacturing, 1*, 3–17.

Borea, M.D. & Wang, B. (2007). Redesign methodology for developing environmentally conscious products. *International Journal of Production Research, 45*, 4057–4072.

Broom, S. (2015). *Sustainable manufacturing minimizes environmental impact while improving bottom line.* Retrieved from http://www.trade.gov/press/publications/newsletters/ita_1008/sustainable-mfg_1008.asp.

Dreher, J., Lawler, M., Stewart, J., Strasorier, G. & Thorne, M. (2009). *MIT Sloan School of Management report: General Motors metrics for sustainable manufacturing.* Cambridge, MA: MIT Sloan School.

Elkington, J. (1998). *Cannibals with forks: The triple bottom line of 21st century business.* Stony Creek, CT: New Society Publishers.

EuroStat. (1999). *Towards environmental pressure indicators for the European Union (EU): An EU report.* Brussels, Belgium: Author.

Feng, S.C., Joung, C.B. & Li, G.(2010, May 19–21). Proceedings of the 17th CIRP International Conference on Life Cycle Engineering: *Development overview of sustainable manufacturing metrics.* Hefei, China: CIRP International.

Ford. (2007). *Product sustainability index.* Retrieved from http://corporate.ford.com/microsites/sustainability-report-2013-14/doc/sr13-ford-psi.pdf.

Glavic, P. & Lukman, R. (2007). Review of sustainability terms and their definitions. *Journal of Cleaner Production, 15,* 1875–1885.

Grote, C.A., Jones, J.M., Blount, G.N., Goodyer, J. & Shayler, M. (2007). An approach to the EUP directive and the application of the economic eco-design for complex products. *International Journal of Production Research, 45,* 4099–4117.

Harland, J., Reichelt, T. &Yao, M. (2008, May 19–22). Proceedings of the IEEE Symposium on Electronics and the Environment: *Environmental sustainability in the semiconductor industry* (pp. 1–6). San Francisco, CA: IEEE.

Harms, R., Fleschutz, T. & Seliger, G. (2008). Life cycle management of production facilities using semantic web technologies. *CIRP Annals - Manufacturing Technology, 59,* 45– 48.

Hawken, P. (2007). *Blessed unrest: How the largest movement in the world came into being and why no one saw it coming.* New York: Viking Press.

Hutchins, M.J., Gierke, J.S. & Sutherland, J.W. (2010). Development of a framework and indicators for societal sustainability in support of manufacturing enterprise decisions. *Transactions of the North American Manufacturing Research Institution of SME, 38,* 759–766.

International Energy Agency (IEA). (2007). Tracking industrial energy efficiency and CO2 emissions. Paris: OECD/IEA.

International Standards Organization (ISO). (2006). ISO 14040: Environmental management-life cycle assessment-principle and framework. Geneva, Switzerland: Author.

Johnson, D.D. & Srivastava, R. (2008, November 22–25). Proceedings from the 39th Annual Meeting of the Decision Sciences Institute: *Design for sustainability: Product development tools and life cycle economics* (pp. 1711–1716). Baltimore, MD: Decision Sciences Institute.

Kaebernick, H., Kara, S. & Sun, M. (2003). Sustainable product development and manufacturing by considering environmental requirements. *Robotics and Computer Integrated Manufacturing, 19*, 461–468.

Karlsson, R.& Luttropp, C.(2006). Eco-design: What's happening—an overview of the subject area of eco-design and the papers in this special issue. *Journal of Cleaner Production, 14*, 1291–1298.

Maxwell, D., Sheate, W. & van der Volst, R. (2006). Functional and systems aspects of the sustainable product and service development approach for industry. *Journal of Cleaner Production, 14*(17), 1466–1479.

McDonough, W. & Braungart, W. (2002). Design for the triple top line: New tools for sustainable commerce. *Corporate Environmental Strategy, 9*, 1711–1716.

Morgan, J.M. & Liker, J.K.(2006*). The Toyota product development system: Integrating people, process, and technology.* New York: Productivity Press.

OECD. (2000). OECD Rome Conference Proceedings Volume I: *Part B— Environmental performance indicators: Frameworks and indicators* (pp. 99–127). Rome: OECD.

OECD. (2009). *Sustainable manufacturing and eco-innovation: Framework, practices and measurement.* Retrieved from http://www.oecd.org/innovation/inno/43423689.pdf.

Parris, T.M. & Kates, R.W. (2005) Characterizing and measuring sustainable development. *Annual Review of Environment and Resources, 28*, 559–586.

Rosen, M.A. (2002). Energy efficiency and sustainable development. *International Journal of Global Energy Issues, 17*, 23–34.

Rosen, M.A. & Kishawy, H.A. (2012). Sustainable manufacturing and design: Concepts, practices and needs. *Sustainability, 4*, 154–174.

Sarkis, J. (2001) Manufacturing's role in corporate environmental sustainability: Concerns for the new millennium. *International Journal of Operations & Production Management, 21*, 666–686.

Smith, C. & Rees, G. (1998). *Economic development* (2nd ed.). Basingstoke, UK: Macmillan.

Stokes, S. (2009). *Get ready for green 2.0.* Düsseldorf, Germany: AMR Research.

United Nations Committee on Sustainable Development. (2007). *Indicators of sustainable development: Guidelines and methodologies* (3rd ed.). New York: United Nations.

Walmart. (2015). *Sustainability product index.* Retrieved from http://corporate.walmart.com/global-responsibility/environment-sustainability/sustainability-index.

Womack, J.P. & Jones, D.T. (1996). *Lean thinking: Banish waste and create wealth in your corporation* (2nd ed.). New York: Free Press.

World Commission on Environment and Development (WCED) (1987). *Our common future.* Oxford: Oxford University Press.

Questions About Sustainability

1. Explain how you are trying to live sustainably.
2. Explain what you think you could do differently in your daily life that would conserve resources.
3. Discuss current, or proposed national policies that you believe could make our natural resources last longer.
4. Write a personal mission statement that defines how you strive to live sustainability.

Chapter 6

Sustainability, part b

This chapter continues the theme of sustainability by focusing on packaging. In chapter 1, we provided some basic information about packaging, to give you some knowledge and help you think about packaging as an entity on its own. In this chapter, we will learn about some of the things that are being done by users and manufacturers of packaging in their attempts to conserve resources and cause lesser harm to the earth.

The first article deals specifically with sustainable practices in packaging. The second article is titled "Cradle-to-cradle" which is a sustainable practice that considers what happens to the materials used—from the creation of the material, through all the processes to make something of it, and then plans for what will happen to it after its useful life is over, whether it is recycling the material back into something else, reusing the item or package for another purpose, or harvesting the energy stored inside the material. Cradle-to-cradle planning helps designers and engineers make intelligent decisions about the most efficient uses of material, about form, size, and other considerations when making new products and the packages that will hold them.

Sustainable Practices in Packaging

Muhammad Aldosari

Sustainable Practices in Packaging

In the past decades, it has been witnessed that there has been an increasing number of customers who want green products. The pressure has also come from environmental stakeholders and the local communities that have turned sustainability into quite a corporate imperative. In various sectors of the economy, key companies have taken the initiative to ensure that they can reduce all the adverse environmental impact they have made. In turn, they are able to enhance their green image. In fact, the many competing brands are working hard to ensure that they can address these concerns, especially in the fast food industry. The world of corporate sustainability is complicated and the progress has continued, showing results. For example, in some instances, the efforts that have been witnessed in the fast food industry have led to great improvement in addressing certain environmental concerns. In other cases, the changes have led to more than just green washing. This paper will provide an analysis of the ways in which the fast food industry engages in sustainable packaging practices.

Sustainable packaging seeks to create an environment whereby all packaging is sourced responsibly. Furthermore, products are designed to be safe and effective during all the stages of their life cycle, and thus they are able to meet all the market criteria for cost and performance. Sustainable practices ensure that renewable energy is used, and upon being used, it is recycled in the most efficient manner, which will provide valuable resources for subsequent generations (Verghese & Lewis, 2012). For example, the use of the closed loop system is extremely beneficial when deciding upon appropriate packaging material.

Sustainable packaging is beneficial since it is healthy and safe for communities and individuals through all phases of the life cycle. It even meets the market criteria for cost and performance. Furthermore, it is often recycled, transported, manufactured and sourced using the renewable energy. In addition, sustainable packaging optimizes the use of the recycled and renewable source materials. It is also known for manufacturing using best practices and clean production technologies (Verghese & Lewis, 2012). In fact, in some manufacturing plants, the packaging is made from materials that are healthy through the whole life cycle. It means that this practice is meant to ensure that optimization of energy and materials takes place.

At least eight main indicators that usually serve as measures in the world are leading fast food practices. First, the industry is embracing leadership on sustainability, and this is the most vital step taken by fast food or restaurant chains. In turn, they are able to make strong commitments adopting environmental packaging policies and sustainability. Once a strong policy exists, it will be possible to buy-in; this will be from the top of the leadership until the bottom, and real quantifiable changes can take place. Most times, when there are organizational changes, successful implementation of sustainable packaging practices and goals will depend on the support of the management. The required approach focuses on integration, takes place throughout the corporate strategy, and will include the boxes, pails, and buckets (Verghese & Lewis, 2012).

The supply chain and full life cycle approach is another sustainable practice in packaging. It has been seen that prudent decisions often originate from the understanding that people have regarding the environmental impact that a product will have. Through the use of science, it is possible to gain an understanding of the materials that are used, and how their environmental impact can be reduced (Verghese & Lewis, 2012). For example, using the Life Cycle Assessment, data can be obtained, and used for providing support for decision making through having a comprehensive approach towards sustainable packaging. It is extremely crucial that the assessment fits within all the sectors of the supply chain in order to uncover the full impacts of the decisions as well as the hidden areas of opportunity that will lead to improvement.

Reducing the overall packaging and increasing efficiency is another best practice in sustainable packaging. The first step that a company needs to take in order to reduce the impacts associated with packaging is that it has to reduce the overall amount of packaging that it uses. The packaging should be designed in such a way that it can minimize the masses of materials that are required to achieve specified levels of functionality. It means that when less of the material is consumed, there will be less packaging that will be used. In turn, this can lead to huge amounts of savings for a company (Verghese & Lewis, 2012). Furthermore, the preferred packaging ought to be physically designed in such a way that it can optimize resource and material productivity through efficiency and light weighting.

Increased use of the recycled fiber is another sustainable practice in packaging. With some effort, it is possible to ensure that there is a reduction of the usage of packaging. It means that it will be almost impossible to have a total elimination. It is recommended that usage of recycled fiber be increased in the paper packaging. Recycled fiber is known to decrease carbon emissions and forest destruction, and even energy, water, and chemical use. In fact, some great strides have been made because of some low-hanging napkins and fruit like bags, and this is quite commendable. The challenge usually emerges when companies have to increase their recycled fiber (Verghese & Lewis, 2012).

The other notable approach is eliminating the paper that originates from controversial forestry practices. It is vital to reduce the overall use of the packaging and increase the recycled fiber content. This will ensure that there is packaging sustainability, as the fiber can continue to come from the forests. It means that companies should be able to understand the supply chain, and thus eliminate the paper that comes from controversial environmentally destructive sources. People should be made aware that paper packaging should not come from the logging of endangered forests. It should also not come from the conversion of the natural forests to plantations, draining and ditching of wetlands, and large scale clear cutting. At the moment, it seems that the best way of ensuring all these issues are taken care of is through utilizing the Forest Stewardship Council certified paper. It is the only known certification that is endorsed by those in the environmental community (Verghese & Lewis, 2012). However, it is not sufficient to claim that the Sustainable Forestry Initiative certification will be sufficient to ensure that companies source from responsible forestry. In turn, this will allow all the destructive practices to be regarded as being sustainable.

Increase in in-store recovery and recycling is another approach towards achieving sustainable packaging. It will not matter how the recyclable packages and products are if they are not recycled, especially when they are in the hands of consumers. The leaders of companies need to ensure that they increase their in-store recycling as well as other standard reduced waste disposal practices. They also need to ensure that they encourage their consumers to recycle all the take-away packaging that they have (Boylston, 2009). For example, if some leaders decide to implement the beneficial management practices, they will surely succeed. They will be able to ensure that people are informed about the reuse and recycling practices that can be achieved. It means that a company can seek to achieve continuous improvement.

Eliminating toxic labels and inks is another way of increasing sustainable practices in packaging. In the recent times, scientific research has highlighted that some health impacts are associated with the dyes, coatings, and inks that come from the food packaging. The elderly and children are the ones that are likely to be affected by the toxic inks. Some simple solutions that are present tend to utilize the various branded packaging and this decreases the potential toxicity of the various materials and pigments used for printing and dying (Boylston, 2009). Furthermore, poor choices have the ability of affecting the recyclability of packaging. Therefore, the health and environmental issues can be addressed if the strategies that include de-dying, water-based dying, and soy ink dying are incorporated.

It is also important to reduce the carbon footprints in order to promote sustainability in packaging. Through reducing the overall usage of packaging, it will be possible to increase the use of recycled fiber. Furthermore, it is recommended that the packaging that originates in destructive forest practices are known. It means that some step towards reducing the carbon footprint must be followed. It should be known that it is also advisable to transport the packaging to the restaurants and this can have a positive impact on the green bottom line. The changes in weight, size, and composition of packaging affect the efficiency of shipping (Boylston, 2009). In turn, this reduces the amounts of carbon dioxide that is released, especially during the transit period. The improved operational choices that include the selecting space are effective pallet configurations. They are able to use the automatic palletizers and thus they choose the fuel-efficient modes of transport that can help in reducing carbon dioxide emissions.

Current best practices in sustainable packaging are key indicators that need to be known. The fast food industry must be willing to embrace the sustainability changes if they are to benefit. Corporate leadership on matters concerning sustainability has to be embraced. It has been established that sustainability is still not a key corporate function in many boards. A lot of commitment is needed, since money and time is usually invested, and this is used in developing the environmentally friendly paper packaging policies. Thereafter, this creates stakeholder partnerships that assist in achieving the sustainable goals (Boylston, 2009). Starbucks is one such company that has made it a goal to make sure that hundred percent of all its cups is recyclable and reusable by the year 2015. The company has gone as far as bringing together the diverse stakeholders so that they can find solutions that will make the cups that are used for hot beverages more recyclable.

The efforts to ensure that all the various representatives engage in sustainability practices are extremely commendable. If this action is taken seriously, the plastic and paper cup value chain will improve quite drastically. It will include stakeholders such as academic experts,

NGOs, beverage business recyclers, cup retailers and manufacturers, raw material suppliers, and municipalities. A number of examples of corporate leadership concerning sustainability exists, and they include the CEOs that initiate the industry paper packaging policy on environmental matters. In turn, this has the ability of increasing the use of the recycled fiber because the CEOs act as the spokespeople for the existing policy. Secondly, the investment in the multi-stakeholder efforts such as those of the Paper Recovery Alliance needs to be prioritized. They are involved in the creation of effective solutions that will be useful in the processing and recovering of the food service packaging used paper such as those of Yum, Tim Horton, and Starbucks, brands. Thirdly, the CEOs should be members of the Sustainable Packaging Coalition as well as other notable environmental packaging forums (Jedlicka, 2008).

The use of a supply chain and full life cycle approach is another key indicator of the current best practices that are present in sustainable packaging. The best available data that is obtained from the life cycle analysis must be the most suitable for packaging material choices. In turn, this ensures that the most effective decisions concerning the issue of sustainable packaging can be made. Certain leadership companies continue to show that they are committed to thinking in a broad perspective, and thus they invest in the resources that exist in the LCA and are warranted (Jedlicka, 2008). The data obtained from the LCA process tends to allow the companies to make some big breaks that come about from past practices. One such example is when companies use the definitive science of recyclability of materials and resource efficiency. Currently, Starbucks no longer uses the standard PET cups, as it now uses the ones that are polypropylene based. The clear, publically accessible guidelines such as those present in the Starbucks' Supplier Social Responsibility Standards and the McDonalds Environmental Scorecard are extremely crucial. They encourage healthy competition among the various supply chains so that they can meet the ever-rising standards of sustainability.

Other examples that come about from using the supply chain and full life cycle approach must also be known. They include the

107

implementation of the impacts of the supply chain and the use of a life cycle approach when trying to consider the various alternatives that are present in all packaging decisions. They also include requesting the suppliers to always provide the needed information on material reduction and carbon footprint. It means working together with other suppliers that work on the end-of-life, and mostly on specific issues. For example, it has been identified that coating is a known key issue, and even constituted work that was conducted together with the supplier in order to come up with a solution (Jedlicka, 2008).

Increasing efficiency and reducing the overall packaging is another recommended practice that will increase sustainability in packaging. Light-weighting and right sizing are two important themes in creating packaging efficiency. In fact, investments are known to pay off together with transportation benefits and reduced costs. A leader such as McDonalds has found certain ways of reconstructing and reimaging the basic components that are paper-based. They are in turn able to reduce the amount of fiber that is used in the various packaging products. It is possible through incorporating the design elements such as corrugation and fluting in order to provide the needed strength that will be made with the paper grades that are lightweight. Furthermore, the small strategic tucks and nips such as the tray napkins and liners are known to reduce the use of fiber (Jedlicka, 2008).

It is possible to increase efficiency and reduce overall packaging through a number of ways. First, re-engineering the corrugated boxes can lead to savings of up to 2 million pounds of the corrugated materials. Secondly, this can be done through reducing the amount of paper fiber commonly used in pizza boxes. In fact, based on some evidence, in the first decade of the 21st century as much as 15 percent reduction was seen. Thirdly, reducing paper that is used in napkins by 21 percent can help. Fourthly, companies need to change their design such as through reducing the sizes of the tray liners that are used to put buns. Reducing the size of the tray by 10 cm can save 84 tons of paper. Some notable companies that make use of the sustainability practices include Subway, Pizza Hut, and McDonalds (Verghese & Lewis, 2012).

Increasing the use of recycled fiber is another key indicator of best practices in sustainable packaging. It is recommended that when addressing the issue of packaging, the origin should be known. In turn, it becomes easy to drastically increase the amount of recycled fibers that are in the different packaging. A number of companies earlier on begun with the low-hanging fruit like napkins and bags. Many of the leaders in this industry have now decided to dig deep into the boxes and cups as well as other containers (Jedlicka, 2008). Currently, Starbucks continues to work hard so that it can overcome the prejudices associated with the old brands. The regulatory grey areas continue to bring to the market about 10 percent of the post-consumer recycled cups used for holding the hot beverages. Recycled fiber use can be increased through making napkins from hundred percent recycled fibers and materials of which 90 percent are post-consumer. Furthermore, the lunch boxes used in the catering industry should be made from the 100 percent recycled paperboard of which 35 percent are post-consumer. It is also advisable to switch to hundred percent recycled bags that are post-consumer. Lastly, all the corrugated shipping boxes need to have a minimum of 35 percent post-consumer, that is, recycled content.

Eliminating paper that originates from controversial forestry practices is another sustainable practice in packaging. Even without having to maximize the recycled fiber, one smart and unique method a company can use to reduce the impacts associated with the packaging on the forest is using certified products. It ensures that the sourcing comes from a working forest landscape that is responsibly managed. Some additional leadership is even derived from the companies that choose to recognize the notable Sustainable Forestry Initiative. It focuses on eliminating and green washing the fiber from the prevailing value chain. In the year 2011, McDonalds introduced the Sustainable Land Use Management Commitment and it had an important function. It means that the company wanted its suppliers to be informed about McDonaldtant function.

It means thfied and transparent FSC when supplying the various forest products. The company would take special care when eliminating the sourcing that comes from the natural lands, protected forests, and are later on converted to tree plantations (Verghese & Lewis, 2012).

The practice of eliminating paper that originates from the controversial forestry practices is unique. It includes using cup and paperboard that comes from sources that have been certified by the FSC. At least 80 percent of the board and paper should be from the recycled sustainable resources or be recycled. Furthermore, 61 percent of the virgin board and paper should also come from the certified sustainable resources. It is also vital to avoid the SFI at times since its use is considered as being a liability that prevails in the industry, and Starbucks is making progress with regard to this issue (Verghese & Lewis, 2012).

Increasing the in-store recovery and recycling is another known and popular sustainable practice in packaging. However, some challenges exist concerning recycling, particularly in the fast food industry. The packaging usually ends up in the customers' hands, and thus a company cannot have direct control over the company. Furthermore, the food packaging tends to remain an issue, as many recyclers have not accepted it yet. The leaders in the various sectors continue to work hard so that they can address all the problems simultaneously. In Yum Yum!, Tim Hortons, and Starbucks, the issue of collaborating in any industry that deals with the food service packaging is considered as being part of a Recovered Paper Alliance. It came about in order to develop all the needed collections of infrastructure (Verghese & Lewis, 2012). In addition, it could even expand to the various uses in the market such as recovering fiber. It could also be helpful for educating consumers on how they can increase their recovery and even partner with the government officials to come up with suitable public policy.

Public commitment and corporate leadership needs to prevail, as this will ensure that there will be long-term success. The leadership at the company should educate all the other employees on the benefits of sustainable packaging, and even come up with some environmental packaging policies. The policies should also have certain commitments and goals that will help achieve success. Furthermore, the various approaches should be focused upon consistently so that they become relevant at all times (Jedlicka, 2008). It means that in the end, sustainable practices will be the norm in almost all companies.

In conclusion, it is evident that adopting the various sustainable practices in packaging is extremely vital, and this is precisely the reason that these practices have been embraced all over the world in various industries. Companies need to have long-term and short-term goals that will enable them to realize their sustainability goals and efforts. All the products that are used in packaging need to be recycled so that they do not destroy the environment. All involved stakeholders must ensure that they play an active role in making sure that only sustainable packaging is used. Packaging that makes use of sustainable practices will undoubtedly provide many benefits to all the concerned parties.

References

Boylston, S. (2009). *Designing sustainable packaging*. Boston: Laurence King Publishing.

Jedlicka, W. (2008). *Packaging sustainability: Tools, systems and strategies for innovative package design*. New York: Wiley.

Verghese., K. & Lewis, H. (2012). *Packaging for sustainability*. New York: Springer.

Chapter 7

Recycling

In the preceding two chapters we discussed sustainability, which is an attempt to make resources last so they do not run out. There are many ways to work toward sustainability. Most people are now aware of what we call the three Rs—reduce, reuse, recycle—but not with the fourth R, recover, which refers to capturing the energy from materials by burning them for fuel.

The first R, reduce, is something people can do by trying to use less of things, bringing their own bags to the store to carry purchases home, or buying things with less harmful ingredients. Packaging engineers are continually trying to reduce the amount of materials used, the amount of non-recyclable materials used, and the amount of toxic elements in inks and adhesives in the packages they design.

The second R, reuse, is something consumers can do. If you get items in a plastic bag at the store, reuse it to carry things later, or use it to hold trash instead of buying new trash bags. Spray bottles can be re-used to hold other liquids that you want to spray onto something. Often, packages are designed with reuse in mind. Many can be used several times before needing to be replaced. The third R, recycle, is what we are discussing in this chapter. Recycling seems to be the one R that people are well aware of, conscious of whether they are practicing recycling or not. Perhaps it is because it makes people feel like they have done some-thing really meaningful if they dump their used packages into a recycle bin instead of a landfill bin. Whatever the reason, recycling is indeed important and encouraging this practice saves a significant amount of natural resources when these recycled materials are reformed into new packages or other products.

In this chapter, five authors will look at different aspects of recycling.

Paper Packaging and Recycling

Abdulmajeed Alhawsawi

On Environmental Awareness and Preference to Paper

Nowadays expanded polystyrene (EPS)—typically called styrofoam, a registered trademark of Dow Chemical Company—is one of the most preferred packaging used for food by many restaurants and different establishments all around the world because it is very light weight and easy to carry. It also looks clean, is sterile, and very hygienic—therefore more popularly used than other containers. Styrofoam is an oil-based container also known as extruded polystyrene, coming in different forms such as bowls, plates, trays, and cups. It is often used as a disposable container made for one-time use only and because it is odorless, most restaurants, food preparers, food servers and packagers prefer this as a container for food.

Though EPS containers seem to be the best possible way of storing and packing food, they pose some health hazards that can be very dangerous to humans. When EPS is used to store food for a very long time or is heated, the styrene in this packaging can go into the foods, which can be dangerous to health because it is a carcinogen. However, even with the health hazards presented by this type of packaging, food businesses still prefer to use it as food containers, more compared to other types of packaging because it is cheap and can be easily found in groceries and supermarkets. Food businesses and most people using EPS would just have to make sure that these containers are used only once and have to try to avoid heating food in it, in order to avoid the hazards brought on by this type of packaging.

Apart from food containers, EPS is also used in industrial packaging. The foam can be used to absorb shock for items that are being shipped. It is flexible, strong, and can be used as a waterproof container

115

for fragile items. And since it is lightweight it would cost a company less money for items that they ship. Because of the many uses of EPS, many industries have preferred this over other type of materials for packaging. But unfortunately, not only is this material very hazardous to human health, but also very dangerous to the environment. Its components are very hard to decompose, even in a hundred years. This is where the risk lies, because eventually, they would end up occupying almost all landfills in just a matter of time. That is why recycling is encouraged in industries that use Styrofoam. However, because it is bulky and light-weight, it may cost companies a large amount of money to recycle it. So in order to resolve this problem they are instead encouraged to put their polystyrene materials into other uses, which in return can further help the companies, because using these materials again would save them money and it can be used on other products, which can mean additional profit for them and at the same time they will have participated in help-ing the environment through recycling.

EPS has been used in different industries for a very long time but until today the dangers and hazards of this material have not been widely known. The public know little of what it can cause to the body and to the environment and even if many industries know this information, they would still rather use it than change to other materials such as paper, simply because other materials cost more and lacks durability, compared to Styrofoam. There should be authorities that should handle these sit-uations and make sure that there is awareness within the public of the dangers EPS can cause. Food and other industries are willing to change from EPS to other materials if there would be cheaper alternatives. If they are forced to change their packaging they would have no option but to increase the price of the food they serve and other products they offer, which in turn can cause several repercussions in the market and even-tually affect the economy. So until there is a cheaper alternative, these companies and businesses would continue to use this material.

Even if it is very hard to totally replace EPS or eradicate its exis-tence from the environment, there are still ways in which to control it in order to lessen its damage to us and to the environment. Recycling alone would not provide the solution of EPS's continued transformation into an overwhelming waste that endangers everyone. Apart from recycling, some people have found ways of putting EPS or polystyrene into other uses that can help reduce the wastes caused by these materials. One innovative solution for EPS was created in a theatre in Austin, Texas, wherein they used polystyrene together with cement and concrete for the wall of the theatre. It was highly commended by many, particularly environmentalist groups because aside from it being able to help with the environment, it ensures durability— even more compared to the normal materials used in walls and buildings—because it can withstand high winds, tolerate high intensity earthquakes, is a good insulator, is energy efficient, indestructible, fireproof and because of its EPS composition, it is able to keep buildings free from pests such as termites.

Another alternative use for EPS was developed in Canada where they put polystyrene foam between two sheets of plywood, calling it Structural Insulated Panels (SIPs). Counting on the fact that this material is a good insulator, the have been using it with fewer amounts of wood to make different kinds of structures. And though the structures use less wood, they still can handle heavy loads. The EPS is also good in preserving heat inside the house or structure because it prevents air movement, which preserves heat and controls humidity. The only downside to these alternative uses of EPS is its cost. The cost of using this material is greater than using ordinary materials but that can easily be remedied, as long as the materials are in constant supply and the demands are big. Eventually, such conditions will decrease the costs. All of these alternatives should be encouraged because they have a great potential for immensely helping the environment and though they might cost more than the conventional ways, the benefits still outweigh the costs (Khalid, Moorthy, Saad, 2012).

On Organizations and Pursuit to Paper Recycling

Because of the volume of goods being produced every day, there is an immediate need for a Life Cycle Assessment (LCA) system of packaging. LCA is very important in packaging because it makes sure that the packaging does not endanger the environment, nor the products it holds, whether perishable or not. Packaging under LCA can be found in different forms, from paper to plastic—such as trays, plates, and bowls, and it was found out that adding high-impact polystyrene to the expanded polystyrene protective buffers can lessen the danger these packaging can cause to the environment in a more effective way. And on the issue of reducing the dangers of packaging on the environment, there have been some innovations and technologies that were developed precisely for that purpose. New innovative packaging have been developed that are environment-friendly. Some examples of these new packaging are those made of synthetic and starch-based biodegradable polymers, polyesters, and active and intelligent packaging materials.

Though the effect of having this LCA system of packaging can benefit the environment immensely, it is not fully practiced by all industries. Its success is not as much as expected, though it shows great potential in helping the environment. One reason why it has limited success is due to small demands of recycles, and that the additional costs involved in collecting, sorting, and cleanings for packaging are too high for it to be a success and that its costs outweigh the benefits. In some parts of the world LCA regulation was implemented, though it was not much of a success. For instance in Germany most industries did continue to be with the LCA system of packaging because of its high cost. In Norway and the UK, the LCA regulation had very little effect on the packaging system of their industries. But in Poland there has been evidence that they began producing sustainable packaging little by little, which would eventually convince other industries to change to this kind of packaging system. It may not have worked instantly, contrary to expectations, but though slowly, this regulation would still make its mark when it is integrated with the proper social, economic, and environmental approach together with a more efficient and recyclable packaging designs (Lee, Xu, 2005).

On Paper Use and Reuse

Synthetic materials such as polystyrene foam used in different industries around the world puts a great stress on the environment. One possible solution that many see is the use of paper, since it decompose more
easily than EPS, and since they are more eco-friendly. But most people
do not know that excessive use of paper can also cause severe damage
to the environment. Overconsumption of paper will inevitably lead to
pollution and waste generation. This is the time that the world should be
more conservative regarding its consumption of paper, since there are
more trees being cut than being planted, which means the supply cannot
meet the overwhelming demands for paper. That is why it is important
to promote technologies and innovations that could help reduce the consumption of paper.

The key in lessening consumption of paper is recycling and reusing. Recycled materials are used as a fiber supply for making paper,
providing approximately 38 percent of the material for paper. But there
are still other ways of increasing this percentage and also there are other
products that paper can be reused to, before being recycled. For example, many believe that being "eco-efficient," they can increase the value
of their products and services and in return for reducing the materials
they use, they spend less energy, generate lesser pollution and lesser
waste. Another possible solution is for shipping industries to sell or
lease reusable containers to their customers. Also, some companies can
sell information transmittal or communications services than reams and
reams of paper and copy machines. With these strategies, many companies and industries would discover that using more materials would
not earn them more money; it is more profitable for them to reduce the
materials they use while providing better services and at the same time
helping save the environment (Abramovitz, Mattoon, Peterson, 1999).

With increasing awareness about paper in solid wastes, there has
been great emphasis on recycling in order to reduce the large volumes

of wastes. There are now several resource conservation methods that have been developed in order to address the problems of wastes. Use of wastepaper in different industries had already begun several years ago, but not until people were made aware of the volume of wastes dumped into the environment that they decided to put emphasis on recycling waste papers and eventually using waste paper not only for profit and economic purposes but also as an illustration of a progressive society. This is how participating countries have been successfully motivating others to practicing this. This has also helped to increase the demand of wastepaper, thus increasing wastepaper collection and improving the wastepaper industry and waste paper use by paper and board industries (McKinney, 1995).

On Green Companies and Paper Packaging

Due to the overwhelming impact of consumption practices of different industries and businesses on the environment, green issues have been raised. The pressure on these industries to go green is because of continuous consumption, marketing, manufacturing, processing, discarding, and polluting that greatly damages the environment. This issue is no longer a local concern but had made its way into a global scale, which is why governments, legislators, and environmental agencies demand that these industries change their system and convert to greener processes in order to help save the planet. Though some companies are truly concerned about the environment, some others did not convert to green for this reason. Most companies agreed to go green because of the pressure they received from government legislations and the pressure from environment activists, NGOs, customers, and stakeholders. And also because by going green, they realized that the company would actually save more money, improve their corporate image, gain approval or benefit from stakeholders, such as those in the financial markets. Their actions may merely be superficial, done in order to secure business gains and not for any overwhelming desire to save the environment.

120

There are also companies that are labeled "low impact organization" or environmental leaders but no matter how environmentally advanced a company is tagged to be, there is no real evidence to validate such claims or to check if they are really doing what they say they are practicing. Companies are not obligated to disclose every detail of their practices and operations so they can say whatever they claim they do for the environment but there is no proof except for the corporate reports that they release, which only benefits them further. This means that companies can express their goals and targets about the environment but that does not necessarily mean that they are able to achieve or meet those goals. And with regards to the green practices, whether for a big or a small industry, there is still a big difference between the environmental impact size and mitigating the overall impact. Most companies would focus on one impact but would not deal with the other impacts on the entire supply chain. For example, marketing hides behind the idea of greening, though it continues to be responsible in using massive amounts of resources and creating equally large amount of wastes.

Greening is an ideal way of saving what is left of our environment, and giving it time to recover from abuse and exploitation done until now. Some may declare themselves environmental leaders but their practices are far from what is expected from them or far from what they claim to be. There are many industries exploiting this serious and much needed ideal for personal gain through communications, policies, and reports behind which their only intention is to improve their image, thus improving their profit. Hopefully these green companies would soon revert from their actions and move towards a greener practice not for profit or gain but in order to save the environment from where all of us get our resources, before it becomes too late and gone (Saha, Darnton, 2005).

The Role of Government in Promoting Paper Packaging

Every person has their own responsibility and their own part in preserving and taking care of the environment. And most of all, people who have authority, like the government, should lead others in battling environmental issues. As a leader, the government needs to formulate possible solutions that would be implemented and should be followed by everyone else, and probably if it is successful, other countries and governments would follow and join the advocacy towards a greener environment. For example in Germany and Japan, they implemented a practice wherein they would include in the packaging waste management the board collecting and recycling. With this, tons and tons of boards would be recycled therefore lessening their waste products. And following examples set by Germany and Japan, other countries had also set their own targets for their packaging waste management directive.

Some governments have been directly involved in this advocacy; they have even tried to improve the market using recycled paper that helped in increasing the volume of collected waste papers. An example would be the paper purchased by the US Federal government to be used for their different programs. Other states also created programs similar to the federal government, such as implementing the use of recycled contents for newsprints. And with the voluntary agreement with UK, they implemented creating 40% of the newsprint from recycled fiber (McKinney, 1995).

Germany, Sweden, Finland, and Denmark are some of the countries who continue to create different programs and set different targets that could help lessen the waste and make great use of recycling and reusing. Some governments are directly—and some are indirectly—involved in this advocacy but one things is for sure and that if the present dedication is continued by countries that are directly involved, there would come a time when all countries and government would work together and lead their people towards positive change and towards a more eco-friendly environment (McKinney, 1995).

On Paper as Used in Packaging Worldwide

Paper and paper board is widely used all over the world. It has many uses, some examples of which include newsprint, books, tissues, stationery, photography, money, stamps, and general printing. But one significant use that accounts for 50% of the total production of paper and paperboard is its use for packaging. The number of paper and paperboard used in packaging is very significant in the total market of paper and paper boards because this is where half of the total production of paper in the world goes into. But apart from this fact, another significant factor is that paper and paperboard is the largest packaging material by weight, used all over the world, which signifies the importance of this material in the packaging market worldwide.

Paper and paperboards are widely used around the globe, and because it can be modified into different things, it can successfully meet the different needs of its consumers. Through different process done to paper and paperboards, there are wide varieties of products that can be made from this material based on the choice of fiber, being bleached or non-bleached, mechanically or chemically separated, virgin or recovered fiber. With the help of these processes, paper can be given different thicknesses and grammages, which can address every need of the market. The appearance of the paper can be altered mechanically. Additives done to the paper during stock-preparation stage can give the paper special properties. Coatings can also be added to the paper, giving it either a dry or a smooth appearance and this feature would have significant effect during printing and conversion and all of these characteristics would create different types of packaging materials (Kirwan, 2008).

On Ethics and Industrial Design in Relation to Paper Packaging

Through the course of history society had gone through technological and industrial revolution. These changes and advancements have generally led to a better life for all humanity. But with these advancements, comes great impact on the environment, such as climate change, degradation of the ecosystem, and other phenomena brought about by development. Because of these consequences, there is now great emphasis on a low-carbon society but the problem is properly creating an industrial design for a low-carbon age. A design is related to human activities and can affect every aspect of modern civilization. It can change and guide people into many things, such as formulating and influencing people's ideas, their concepts, and ethics. Because there are many things that can cause temptations and resistance and with the natural characteristics of some people being lack of control, a strong design ethics, construction people, and objects, people and nature of ethical awareness.

On Paper as Used in Packaging Worldwide

Paper and paper board-based packaging are widely used because they are able to meet the criteria for a successful packing—containing a product, protecting the goods from any mechanical damage, protecting the production from deterioration, informing the customer and providing a visual impact—due to its graphical and structural design. Not only that this material has met all the necessary criteria but paper and paperboard-based packaging also meet the criteria at all three levels of packaging. The primary level of packaging is one that is used in a single product unit at the point of sale or use, the best example of this being cartons. The second level of packaging includes those that are used in collections of primary packs grouped for storage and distribution, such as transit trays and cases. And the tertiary level is for big unit loads for distribution in bulk; these are heavy-duty fireboard packaging.

Paper and paper board have successfully met all of these criteria because of its many forms, appearance and properties. That is why it can be developed into a wide variety of package structures that can be used in different types of units at a very low price, making it an ideal material for packaging. Other characteristics for which paper and paper-board is an ideal form of packaging material are its printability, varnishability and capability of being laminated with other materials. Its physical properties make it very flexible and that is why it can be used in different forms and shapes. It can be cut, creased, folded, formed, winded, and glued to other items, making it easy for manufacturers to create different packages with different size and shapes. It can even withstand high range of temperature, from frozen food storage to temperatures of boiling water heated in a microwave oven or conventional oven. Having met the criteria and also at different levels, having different uses and characteristics, being flexible, innovative, and also having the ability to be recycled and reused, paper and paper boards can be considered to be one of the best options as packaging material in different industries (Kirwan, 2008).

References

Abramovitz, J. N., Mattoon, A. T., & Peterson, J. A. (1999). *Paper cuts: Recovering the paper landscape.* Washington, DC: Worldwatch Institute.

McKinney, R. (Ed.). (1995). *Technology of paper recycling.* Springer.

Khalid, K. A. T., Moorthy, R., & Saad, S. (2012). Environmental ethics in governing recycled material Styrofoam for building human habitat. *American Journal of Environmental Sciences, 8*(6), 591.

Kirwan, M. J. (2008). *Paper and paperboard packaging technology.* John Wiley & Sons.

Lee, S. G., & Xu, X. (2005). Design for the environment: Life cycle assessment and sustainable packaging issues. *International Journal of Environmental Technology and Management, 5*(1), 14–41.

Saha, M., & Darnton, G. (2005). Green companies or green con-panies: Are companies really green, or are they pretending to be? *Business and Society Review, 110*(2), 117–157.

Yong-tao, L. (2012). Industrial design ethics in low-carbon age. *International Proceedings of Computer Science & Information Technology, 52.*

Virtanen, Y., & Nilsson, S. (2013). *Environmental impacts of waste paper recycling.* Routledge.

Analysis of Corrugated Box Recycling

Moaed Bokhari

Introduction

Recycling of different products is quite essential for purposes of ensuring that the environment is protected from plastics as well as other harmful products. Through the recycling of corrugated boxes, there is a decrease in the solid waste disposals in the landfills. During the recycling process, corrugated box provides adequate fiber that is reused for purposes of making new corrugated boxes. To this end, less new raw materials are used in the recycling process. It is crucial to note that recycling of corrugated boxes ensure that even the end-users earn revenues because the recovered materials from the old corrugated containers prove to be valuable to paper mills as well as the manufacturers of new corrugated sheets. There is an urgent need for the society to conserve the environment through making smart and responsible choices that encompass recycling of the packaging materials and other wastes.

How to Recycle Corrugated Boxes

It is imperative to note that corrugated boxes are largely used for packaging of electronic products and food products. They are brown in color and they acquire the name "corrugated" from a layer of fluted corrugated paper. All Recycling Facts (2014) asserts, "One reason for the high recycling rate on the corrugated boxes is that recyclers are paid for the sale of the old corrugated containers (OCC) back to the corrugated industry" (Para.8). Before recycling, any contaminants like Styrofoam or plastic wrapping are removed from the corrugated box and in most cases, clean corrugated boxes fetch higher prices. The contaminants have toxic materials that are not suitable for recycling. There are different companies that engage in the process of recycling corrugated boxes

126

and as such, there is the need to ensure that all corrugated boxes lying in the compound are taken for recycling. Corrugated boxes treated with plastic extrusions—waxy or even laminated—are not easy to recycle. The recycling process starts by sorting the various corrugated boxes on site. The clean corrugated boxes are then taken to the next step. The OCC that are clean are compacted, then baled for purposes of ensuring that there is efficiency in terms of storage and handling. The baled OCC are later transported to the paper mill and fed into a repulper that is filled with water. The function of the repulper, which is a giant blender, is to aid in reducing the OCC into fiber and water (All Recycling Facts, 2004, para.17). The slurry from the repulper passes through several tanks to ensure that all contaminants are removed. In this case, contaminants may float above and be sieved away or they may sink to the bottom of the tanks and be removed. Water is left to drain away and when the fiber layers passes through heat, moisture is removed making the dried fiber layer as paper. The single-facer machine is responsible for softening the recycled paper with steam and finally, starch adhesive is commonly used for purposes of sticking the fluting to the liners to form a complete corrugated box.

It is not necessary to recycle the old corrugated containers by passing them through different processes with the view to creating new corrugated boxes. There are new ways of recycling the containers as well. Unused boxes should be dismantled and laid flat to avoid the boxes from taking up much space. Before commencing on how to recycle the boxes, there is the need to ensure that the project is planned properly. Dimensions based on sketches of what one wants to design helps in utilizing the corrugated boxes fully. Bayan (2015) posits that craft knife or box cutter are used in cutting up thick cardboard as scissors are often inadequate in cutting such cardboards (para.2). Corrugated boxes can be used in making an art frame. This is because the flat and rigid surface of the cardboard panel eases the work of cutting and designing the appropriate final product. Anyone can undertake recycling of the corrugated box by following easy steps in designing various art frames within the household.

When one is making an art frame, he or she should stack strips from the corrugated boxes and then glue them together in order to create the frame borders. The borders are then wrapped and the craft paper is glued for purposes of giving the artwork a natural look. The piece of art can then be hanged on the wall. Further, making a trash bin is made in the same way as the art frame, only that the desired shapes and sizes are cut according to the user's wishes. Corrugated boxes can also be used in making mini-shelf and divider for the sock drawer, besides being used as file organizers. An individual can utilize his or her own imagination to design different cardboard crafts and décor items for the household (Bayan, 2015, para.8). The use of color depends on an individual and thus, the best shapes will be dependent on the imagination of an individual. Despite the absence of recycling processes that converts the OCC into a new cardboard box, they can be utilized to decorate the house instead of being thrown in the pits.

People are often unaware of where to take their cardboard boxes after using them for different functions. It is advisable not to take the boards to the litter since they are biodegradable and as such, they should be taken to the nearest recycling centre. Tapes and labels can be left on the cardboard since the workers at the recycling center have the duty of removing them as well as any contaminants present in the boxes. Earth 911 (2014) asserts that corrugated cardboards can be sold in the market regardless of the medium market demand for the product (para.3). Failure to get a market for the cardboard should not encourage people to litter these boxes across the neighborhoods as that might have adverse effect on the environment. Cardboards are usually sold in bulk and as such, it would be advisable to take used cardboard boxes to the recycling center, as these organizers will be able to sell them in bulk in turn. Recycling of the cardboard is important in the sense that it enhances the conservation of the environment besides making people aware of their responsibilities in terms of going green.

It is quite a herculean task to recycle wet cardboards. This is because wet cardboards can clog automated sorting machines that are utilized in the recycling process and thus fail to yield to getting new cardboard boxes. Wet cardboards should be thrown aside to avert contaminating an entire recycling load of cardboards (Earth 911, 2014, para.4). Corrugated cardboards are recyclable by their very nature. However, pizza boxes are hard to recycle owing to presence of grease and goods. Presence of different contaminants and harmful items has the effect of causing damage to the machines during the recycling period. Caution is important for the safety of the employees, as mishandling of the corrugated boxes can hurt the handlers of the boxes. Before recycling, there is need to ensure that the boxes are fully flattened, as that would aid in the transportation of cardboard boxes to the recycling institution. The whole process of recycling is easy and quick; thus, there is need for the society to be made aware of the benefits of recycling in order to encourage more recycling of corrugated boxes.

Process of Recycling

The first process in recycling of the corrugated boxes is making of the pulp. In this process, all the collected corrugated boxes are carefully sorted and send to the paper mills. The boxes are well agitated until they form fiber slurry. In addition, the contaminants that hang on the ropes as well as the chains are removed. Metal falls, while lighter materials like plastics usually float to the top and are easily removed. Korpella (2015) posits, "Recyclers separate corrugated boxes from other paper because paper mills produce different grades of material, depending on what is being recovered" (para.1). There is the need for entities to separate the corrugated boxes with the view to ensuring that no damages are done to the recycled corrugated box, as small glitches can have adverse effects on the entire recycling process. Proper campaign on recycling corrugated materials can have positive impact on the society.

After metal and other contaminants have been removed, the cleaned pulp is taken to the machine where the water is drained through a moving screen, leaving a fiber mat. Heated cylinders are utilized in ensuring that the fiber mat is dry. Before shipping to the box manufacturers, the paper is usually rewound into rolls. In most cases, manufacturers utilize recycled paper for the flat liner sheets besides the corrugated ridged or the fluted medium between the liners in order to make new corrugated boxes (Korpella, 2015, para.5). It is crucial to note that the liner paperboard that is used for purposes of making the corrugated boxes may be a hundred percent but most of it is usually thirty-five percent recycled content. In 2009, corrugated boxes were considered the largest single component that constituted the municipal solid waste. Further, recycled cardboards are instrumental in being used to stop weeds, plant, create home décor projects, as mulch, besides other creative uses for different functions around the household.

Why should Cardboard be Recycled

Cardboard is a readily recyclable material. Waxed, soiled or wet cardboard is usually not recyclable but can be used as compost in various commercial composting operations. Some states have tough laws regarding waste. Recycling works (2015) states, "In Massachusetts, all cardboard, paper, and non-waxed cardboard products are banned from disposal by the Massachusetts Waste Bans" (para.2). With the advancement in economy, businesses as well as different institutions recycle items like cardboard and as such, save money on waste disposal costs. Through recycling, the society is able to conserve valuable resources besides reducing pollution from the production of new materials, and creation of jobs. People may receive revenues from generating large cardboards and selling the same directly to the market. It is important to note that cardboard can be recycled many times without losing its strengths.

Benefits of Recycling

Corrugated cardboard is largely used in making of brown packing boxes and it is usually identified by a layer of cardboard, thus making it a three-layer sandwich of cardboard. The other type of cardboard is the paperboard or chipboard, which consists of a single layer of gray cardboard that is commonly used for making shoeboxes, cereal boxes and other packages. Lallanilla (2015) asserts that cardboard containers that are coated with wax or other substances to provide them with more strength when wet cannot be recycled (para.4). It is imperative for people to check with their collectors or even city governments in order to know how they handle the different types of corrugated boxes for purposes of recycling. Juice containers, milk cartons as well as some boxes that have resin or even wax-coat are not easily recyclable. Presence of these contaminants will have the effect of contaminating other boxes and thus affecting them.

However, it is imperative to note that if a given corrugated box cannot be recycled, it can be used for other purposes around the house. One can compost and use the cardboard in the compost pile. In addition, it can be used to line garden beds or for mulching, as a weed control measure. Further, the box can be utilized for shipping or storage. Advocates assert that recycling of corrugated boxes saves 25% of the required energy that is required for purposes of making new cardboards (Lallanilla, 2015, para.7). Recycling of the corrugated boxes is sustainable as compared to cutting down of trees with the view to making cardboard products. There has been concerns over forest cover and as such, cutting down of trees is not viable, given the recent climatic changes. It is crucial to note that every ton of recycled cardboard saves none cubic yards of landfill space. Recycling is thus the best option to averting environmental degradation, besides ensuring that energy is saved.

Statistics provides a weary report on the use of paper. Leblanc (2015) points out, "According to the Environmental Protection Agency (EPA), nearly 46 million tons of paper and paperboard were recovered in 2011, translating into a recycling rate of around 66 percent" (para.2). These papers are generated from the offices, grocery stores, supermarkets and small convenience stores. There is thus the need to play a crucial role in increasing the number of wastes taken for recycling to avert destruction of the environment. When the public is provided with some of the statistics regarding the usage of cardboard boxes, they may be able to embrace recycling of the materials besides active participation in ensuring that waste that is recyclable is taken to the recycling centers and thus blocking environmental degradation. In addition, the different stores that use these cardboard boxes should also be educated on how to help their consumers embrace the importance of recycling and thus, take their cardboard boxes for recycling once they are through using them.

There exists various benefits when one recycles old corrugated containers instead of discarding them. It is imperative to note that recycling of the OCC has the benefit of conserving energy as well as water usage, apart from averting production of greenhouse gases and different air pollutants such as the total reduced sulfur (TRS), volatile organic chemicals (VOCs) and the hazardous air pollutants (HAP) (Leblanc, 2015, para.4). It is significant to note that through recycling of the cardboards, there is reduced demand for the virgin timber that is derived from felling of trees and thus posing more danger to the already vulnerable environmental conditions of global warming. Recycling helps in reducing the wastes in the landfills and as such, reducing waste tipping fees. Large generators of cardboard boxes usually deal with paper companies while mid-level generators have to deal with the recycling companies. Irrespective of how the OCC is generated, there is the need to recycle them.

The design of corrugated box helps in allowing large amounts of empty air to be present within the cardboard and thus in reducing its weight, which aids in making it much more impact resistant. Over eighty-five percent of all products available in the United States are shipped and packaged using corrugated cardboard. Smith (2015) asserts that because the corrugated cardboard boxes take up large space owing to its weight, it is imperative to keep it from filling up landfills (para.4). Through recycling, less fossil fuel is burnt and as such, there is less pollution that enters the atmosphere. It is important to note that recycling companies do not accept any cardboard that is wet as the water usually breaks down the paper fibers and thus reduces its usefulness. It is for this reason that the cardboard pulp is usually mixed up with fresh pulp during the recycling process. Pollution has become a menace is the recent years and with the world going green, there is the need for different companies to embrace the noble idea of recycling in order to make the world a better place for the posterity.

Reducing, Reusing and Recycling of Cardboard and Brown Paper Bags

Cardboard is usually manufactured from the cellulose fibers, which are largely extracted from trees. It is important to note that making of the pulp that is used in recycling of cardboard helps in creating sulfur dioxide gas that helps to cause acid rain. Recycling of cardboards helps in cutting pollution by half. Paper fibers that are used in the making of corrugated boxes are long and strong and as such, they can be used several times, consequently reducing the need to cut trees down. Valley Community for Recycling Solutions (2015) posits that preserving of the forests has benefits to the rivers and lakes by preventing erosion, enhancing the air that people breathe by removing carbon dioxide, besides adding beauty to the surrounding, and saving animal habitats (para.2). People should reduce the number of cardboard boxes as well as brown paper bags that they use and wherever possible use the smallest sized boxes, totes, or cloth bags for grocery shopping.

Cardboard boxes should be reused for storage, moving, mailing as well as collecting recyclable items. In the process of sorting the different papers that one intends to recycle, there is the need to look for the wavy inner layer that distinguishes corrugated boxes from other boxes. Bales that contain more than 5% of other paper fibers, 1% prohibitive or 10% of moisture are rejected and eventually end up in the landfill (Valley Community for Recycling Solutions, 2015, para.8). Magazines, egg cartons, and paperboards have presence of shorter fibers as well as fillers that serve to decrease the quality of the final products. Further, materials like rocks, glass, Styrofoam, wax and even wet paper have the effect of damaging the recycling equipment, besides causing health concerns and making the final product unusable. It is also imperative to note that presence of moisture in the cardboard has the effect of weakening fibers and causing dangerous conditions during the recycling process. Americans are known to throw away as much wood and paper annually as can be used to heat five million homes for the next 200 years.

Packaging is quite essential for all products and as such, corrugated boxes are used in the supply chain. Most entities focus on reducing the costs of using corrugated boxes. Industrial Asia does not have major softwoods as compared to North America and North Europe as in these nations pulp and papers have been well developed over 100 years. Asia relies heavily in recycled paper as its major source for recycled paper. Ampuja et al. (2014) asserts, "However, each time paper is ground up for recycling, cellulose fibers become shorter, making the paper softer and less strong" (p. 32). An efficient supply chain that moves different products from the supplier to the consumers is quite important. Concerns over damages and environmental impact from the use of different packaging materials have resulted in most companies recycling the corrugated boxes in a bid to lower their overall costs and get the products to consumers in a safe manner.

Some institutions have been encouraging people to be conscious about the recycling of different materials. Institutions like Virginia Tech have academic programs that combine packaging as well as industrial design with the view to equipping young professionals on how to deal with the issue of recycling. Owing to higher logistical costs, companies like Walmart have been striving to reduce their logistical costs. In Asia, lack of recycled corrugate has led to the use of double as well as triple wall boxes for purposes of shipping textiles that require high level protection. The mechanical pulping process practiced in Asia results in more lignin in the final recycled product, as compared to those recycled in Europe and North America (Ampuja et al, 2014, p. 32). Lignin has the effect of making the resulting paper weak besides causing discoloration over a given period. It is for this reason that Asian corrugate is described as rice board or even rice paper by many people in North America.

Conclusion

Recycling is quite an imperative for the society and as such, people should be conscious about how they dispose of corrugated boxes. As we have seen, corrugated boxes can be recycled, courtesy of different recycling companies, and the final product sold to the market. In addition, one can use these corrugated boxes for designing items of home décor. Corrugated boxes can also be used in making mini-shelves, dividers, and file organizers. In the recycling process, there is the need to sort out the different corrugated boxes, with the aim of separating dry boxes that have no contaminants. Any wet materials or contaminants have the effect of causing damages to the final product. Juice containers, milk cartons as well as some boxes that have resin or even wax-coat are not easily recyclable. Through recycling, the society is able to conserve valuable resources besides reducing pollution from the production of new materials and creation of jobs. Through recycling, less fossil fuel is burnt and as such, less pollution enters the atmosphere.

References

All Recycling Facts. (2014). Recycle sign for corrugated boxes. *All Recycling Facts.com*. Retrieved 22 Apr. 2015 from http://www.all-recycling-facts.com/recycle-sign.html

Ampuja, J., White, M. S., Venkatesh, V. G., & Dubey, R. (2014). Packaging: Think inside and outside the box. *Supply Chain Management Review*, *18*(5), 30–39. Retrieved 22 Apr. 2015 from http://eds.b.ebscohost.com/ehost/pdfviewer/pdfviewer?sid=ff8c40ea-070d-44df-ab15-c1b320a4b4f6%40sessionmgr112&vid=0&hid=117

Bayan, R. (2015). How to recycle cardboard boxes. *EHow*. Retrieved 22 Apr. 2015 from http://www.ehow.com/how_2267902_recycle-cardboard-boxes.html

Earth 911. (2014). How to recycle cardboard. *Earth911*. Retrieved 22 Apr. 2015 from http://www.earth911.com/recycling-guide/how-to-recycle-cardboard/

Korpella, R. (2015). How are cardboard boxes recycled? *Sfgate.com.* Retrieved 22 Apr. 2015 from http://homeguides.sfgate.com/cardboard-boxes-recycled-79171.html

Lallanilla, M. (2015). How to recycle cardboard. *Greenliving.com.* Retrieved 22 Apr. 2015 from http://greenliving.about.com/od/recyclingwaste/a/Recycle-Cardboard.htm

Leblanc, R. (2015). Old corrugated cardboard recycling. *Recycling.com.* Retrieved 22 Apr. 2015 from http://recycling.about.com/od/Paper/fl/Old-Corrugated-Cardboard-Recycling.htm

Recycling Works. (2015). Recycling cardboard. *Recycling Works Massachusetts.com.* Retrieved 22 Apr. 2015 from http://recyclingworksma.com/how-to/materials-guidance/recycling-cardboard/

Smith, A. (2015). Facts about recycling corrugated cardboard. *EHow.* Retrieved 22 Apr. 2015 from http://www.ehow.co.uk/info_8602655_recycling-corrugated-cardboard.html

Valley Community for Recycling Solutions (VCRS). (2015). *Recycling cardboard. Valley Recycling.org.* Retrieved 22 Apr. 2015 from http://valleyrecycling.org/what-how-to-recycle/recycling-cardboard/

Aluminum Can Recycling

Abdulrahman Sinnari

Introduction

The sector affiliated to aluminum can has the objective to steadily increase recycled aluminum content in the manufacture of fresh aluminum cans, in a manner to increase environmental performance, reduce costs, and target the aluminum cans recyclability. In order to mount the content of recycled cans, more old scraps are being used in the production of a new product. Aluminum is named as the third most plentiful metal in the earth's crust and also the second most mainly used metal around the globe, owing to its mechanical and physical properties like corrosion resistance, light weight, non-toxicity, manufacturability, and heat conductivity (MacGregor, 2003). In comparison to other metals, aluminum manufacture is in its original stages, and the infiltration of the application of this metal in the community is growing rapidly. The use of aluminum is mainly in the alloyed form. Aluminum forms are produced from different alloy series. Each of the variety has a diverse key alloy, which alters aluminum properties like reactivity, softness, and formability.

Aluminum is a sustainable precious metal that can be recycled as many times as possible.

Businesses have furthered and increased their competitive advantage via the use of aluminum. Aluminum sustainability has led to the creation of jobs and the creation of competitive advantage. Aluminum can be recycled to a new can and get back to the supermarket shelves in 60 days. Cans available are either picked from community drop-offs or curbside pickups. The use of aluminum cans is associated to competitive advantage. Gaining competitive advantage is an aspect that every business would want to achieve so that they can have a better marketing and working position. A good competitive advantage makes business attain a higher market share.

The use of aluminum cans can draw consumer goodwill. The use of aluminum cans make them convenient, easy to use, durable, and light and these aspects make them a favorite aspect across many businesses. The use of aluminum is also the aspect that it offers a better printable surface that accepts paints and coatings, and primary advantage for attractiveness and brand identification of packaging.

Aluminum is a metal that is made from bauxite ore. The ore bauxite comprises mainly of iron oxides, aluminum oxides, and clay. Bauxite is trodden and mixed with canstic soda that removes dirt, leaving a bright white powder that is then frenzied to remove any moisture. The heating leads to the production of au alumna or pure aluminum oxide. Smelting is done on the pure alumna in a large steel furnace to remove oxygen and leave pure molten aluminum. The pure alumna is then poured to f01m ingots. Conversion of aluminum from bauxite is not as easy as recovering the metal or melting the ore (Pohl, Acosta, & Gutierrez, 2007). A high electrical power is passed during the alumna alienating oxygen and leaving the fresh alumna that can be poured off and melted. Small amounts of silicon, magnesium, and manganese are added to the thick alumna to offer it for increased resistance to rust, strength, and better casting properties. Due to the complicated method of converting bauxite to alumna, five tons of bauxite can make about one ton of the alumna.

Once hot, alumna can be rolled along large rollers to create thin sheets necessary for the production of foil and cans. Aluminum foil and aluminum cans are produced from the best quality alumna since only high-quality alumna can be rolled and pressed so thinly and also stay strong. Cans are good for recycling as the recycling expense is less compared to its manufacture from raw materials. The reason aluminum recycling is good is that only 5 % of the energy is wanted to recycle aluminum to rebuild aluminum stuffs from aluminum cans. A ton of recycled aluminum can make a ton of fresh aluminum. In recycling, there lacks waste materials.

Aluminum recycling is also preferred as it leads to the conservation of mineral resources, reduces energy usage from transportation and mining, and reduces waste to landfill. Aluminum recycling leads to environmental and economic importance since many alloys, and pure metals require far less energy to remake as compared to extracting, mining, and smelting. As a matter of fact, twenty cans could be transformed with similar energy equivalent to manufacturing a fresh can from raw materials (MacGregor, 2003). Aluminum aerosols and cans are recycled to help in saving natural resources, but when discarded in landfills, they could take about 500 years to decay.

Recycling is the technique of taking products or materials that are at their life end and transfo1ming them into either a similar product or a secondary material. The method of recycling can be hugely categorized into two different categories, which are closed-loop and open-loop recycling. Recycling of aluminum demands the use of the closed-loop process. The use of the closed-loop process ensures that the same material is processed and made to a similar product (Askeland, Fulay, & Wright, 2011). Decisions regarding whether to use a closed-loop or open-loop model depends on the frequency of recycling the product and the accessibility of data. For products manufactured by the open-loop technique, its products are not similar to the inputs.

Aluminum is a metal that forms about 8 % of the earth's crust and is present in vegetations, most rocks, and animals. Pure aluminum is said to be very reactive and cannot be found as a free metal or element in nature. Aluminum is a metal that is soft and lightweight. There has to be a combination of aluminum with other metals like magnesium, iron, copper, silicon, lithium, zinc, and titanium, for instance, to manufacture a selection of alloys of diverse properties for special purposes. Aluminum has a significant role to fuel-efficient engines for use in transportations.

Aluminum also leads to the construction of low maintenance and corrosion resistance building.

Aluminum is also widely applied in packaging for storage, protection, and preparation of beverages and food. Aluminum can also be rolled to ultra-thin foils that are strong, light, and have an exclusive insulation and barriers capabilities to preserve beverages and food against odor, ultra-violet light, and bacteria. Aluminum packages are tamper-proof, secure, easy to open, hygienic, and recyclable. Almost each aluminum product manufactured commercially can be recycled after its life ends, without losing its metal quality and properties.

Aluminum is a metal that possesses versatile properties that make it easy to mold, cast, roll and extrude to an infinite array of shapes. Aluminum properties range can be found in a lot of commercially existing alloys. An alloy is an element that comprises of two or more metals.

Alloys developed and produced possess certain desirable, specific characteristics like formability, strength, and corrosion resistance (Pohl, Acosta, & Gutierrez, 2007). The logic and composition of the alloys are synchronized by a globally agreed classifications structure or nomenclature for shaped alloys and by several domestic nomenclature formats for alloys casting.

Recycling is a key business operation linked to the aluminum industry. In Canada and America, about 5 million aluminum tons are recycled every year and most of the recycled materials and cans are injected back to the main stream supply. The aspect of aluminum recycling consumes about 92 percent of energy than manufacturing new cans, the aspect of recycling is viewed as an environmental and business win for the sector. Companies across the globe are incorporating sustainability and environmental objectives to their business mission facts. Aluminum is a metal that was discovered in 1820 and was said to be the most plentiful metal on earth.

The use of aluminum since then has been in the manufacture of various items like gutters, aluminum cans, aluminum foils and other items. In America alone, about 100,000 cans are recycled daily. Every recycled can implies that more resources are available at a reduced cost. It mainly takes almost 6 weeks to make, fill, trade, recycle, and then recreate a beverage can. Recycling is a practice that is common and which existed form 1900. During the 1960's the practice became common after aluminum beverage cans became popular with the public. The use of aluminum is popular as it is a strong light weight material that possesses high thermal conductivity.

Recycling of an aluminum can is a process that does not take many days. Within some few days, aluminum can go back to the store shelf after a month. The process of recycling is a very simple process as the can is remolded and turned to a sheet that is then formed to another aluminum can. The recycled aluminum can have just the same qualities and characteristics like the melted can. Aluminum cans are recycled with no deprivation of quality thus making them the perfect products for closed-loop advance to recycling. Among the various aluminum recyclers Novelis who are the world's biggest aluminum can recyclers (Newell, 2009). The recycling process basically has and talces four processes.

Can shredding

The first step begins with talking all the aluminum cans and putting them in the shredder for them to be shredded. Shredding is the process of reducing the size of the aluminum cans so that they can be sorted. Shredding involves cutting the aluminum cans into small sizes for assortment, which is done to separate it from any other material. The shredder produces a 1,000 horsepower, where the aluminum cans are shredded and are later moves on to a double magnetic drum. In the double magnetic dtum, sOlting begins. The drum acts like a separator, where separates any steel from aluminum, and such is kind of steel could have been assorted into the bale.

141

De-coating

The next phase after shredding involves de-coating. The process or phase of de-coating involves the removal of any coat or color or paper from the shredded aluminum cans. The process involves the use of heat, which is mainly done through the use of blowing hot air with temperature of around 550 degrees through the shreds and it is done via a slowly moving protected conveyor (Inches, & Chambers, 2009). The exhaust gases emitted from this phase are first directed through an afterburner and the same gases are used to heat the inward process air through a heat exchanger, decreasing the energy supplies of the system.

Melting

The subsequent method after de-coating is melting. The aluminum shreds are then moved on to the melting furnaces that include stirrers, which develop a vortex in the molten aluminum pool and drags the shreds fast down to the melt. The process is one among the many that requires a lot of heat to melt the aluminum shreds. The process attains high yields and rapid melting rates. The furnaces are equipped with state-of-the-art burner management and burners, and have fuel efficient systems aimed at reducing the degree of energy applied and the effects it's possesses towards the environment (Askeland, Fulay, & Wright, 2011). The furnaces are also said to be equipped with jet stirrers that ensure an even composition and temperature by encouraging metal circulation inside the furnaces. The stirring process has the capability of ensuring the production of the biggest-quality end product.

Aluminum casting

The final phase of recycling is the aluminum casting. The molten metal is then moved to a holding furnace where treatment begins. The molten metal is treated so that impurities can be removed before the molten metal is casted to aluminum. Due to its versatile properties that make it easy to mold, cast, roll and extrude to an infinite array of shapes, aluminum is easily made to different shapes that suits the beverage can. Blocks are made by slanting the furnace and letting the molten metal flow to a casting unit. The metal is diagnosed in a two-stage procedure to eliminate any remaining minute non-metallic gases or particles with metal cleanliness and chemical composition tested on every cast. Aluminum alloys possess certain desirable, specific characteristics like formability, strength, and corrosion resistance. The logic and composition of the alloys are synchronized by a globally agreed classifications structure or nomenclature for shaped alloys and by several domestic nomenclature formats for alloys casting. The aspect of making alloys is to offer aluminum with a more extensive life and also ensure that it does not rust, corrode, or decay easily (Inches, & Chambers, 2009). As the metal streams to the molds, it is frozen by jets of cool water drained though and around the mold base. The aluminum ingot hardens steadily during the casting process that can take up to three hours. The completed 18-ton ingots, every having roughly 1.5 million old cans, are ferried to a mill for rolling to the sheet where aluminum cans manufacturers then create new cans and the entire process commence again.

Aluminum facilities must be regulated and assessed by the environmental agencies of the nations in which they operate and should also meet their strict control measures. The waste gases produced from the four processes should be eliminated from the plant and diagnosed in purpose built emission amenities including hot dust scrubbers ad cold dust collection structures. The use of the cold dust scrubbers is to eliminate the exhaust gases from the conveying and shredding procedures while the hot dust scrubbers manage the case from the melting, de-coating, and holding furnaces. The other materials made from aluminum are also recycled with the same process, but with requirements of various processing and upgrading technologies (Askeland, Fulay, & Wright, 2011).

Advantages of aluminum can recycle. There are several advantages affiliated to the recycling of aluminum cans. The advantages include:

1. Energy saving

    Aluminum recycling consumes about 5 % of the energy necessary to produce alumna from bauxite. Then above aspect indicates that for every pound of aluminum recycled, the aluminum sector saves the energy funds necessary to generate 7.5 kilowatts of electricity. The amount of energy saved by the recycling of about 1.5 billion pounds of alumna could sustain an entire city for about six years (Haggar, 2007).

2. Helps the economy

An aluminum can is one of the most valuable containers for recycling and is the most recycled material in many parts of the world. Aluminum possesses significant market values and goes ahead to offer an economic enticement to recycle it. Locally, if aluminum containers are recycled curbside, they aid in paying for community services. Annually, the aluminum sector pays a lot of money for empty aluminum cans.

3. Helps the environment

For the production of aluminum to continue, the extraction of bauxite has to happen. That requires mining, transportation, and processing. The process of changing the bauxite to alumna requires a lot of energy and a lot of waste materials are produced (Inches, & Chambers, 2009). Recycling helps save from the production of carbon dioxide that would lead to global warming. The other gasses released from bauxite conversions are also cut short due to recycling.

Conclusion

Aluminum, as indicated, is preferred as it leads to the conservation of mineral resources, reduces energy usage from transportation and mining, and reduces waste to landfill. Its abilities of being an exclusive insulation and barriers capabilities to preserve beverages and food against odor, ultra-violet light, and bacteria makes the material necessary for the production of beverage and food materials. Aluminum packages are tamper-proof, secure, easy to open, hygienic, and recyclable. Recycling is the technique of taking products or materials that are at their life end and transforming them into either a similar product or a secondary material.

The technique of recycling can be hugely categorized into two different categories, which are closed-loop and open-loop recycling. The use of aluminum cans make them convenient, easy to use, durable, and light and these aspects make them a favorite aspect of many businesses. The use of aluminum is also the aspect that it offers a better printable surface that accepts paints and coatings, and primary advantage for attractiveness and brand identification of packaging (Gitlitz, 2002).

Recycling should be a process that should be highly recommended to ensure that conservation of the environment and increment of the economy is encouraged. Recycling is a process that also reduces waste materials, which could have been harmful to animals, plants, and people. Each company should be advised to use aluminum as aluminum forms are produced from different alloy series. Each of the variety has a diverse key alloy, which alters aluminum properties like reactivity, softness, and formability.

Each company should strive to use aluminum as it helps a company attain a competitive advantage and this ensures that the company has a better marketing and working position. A good competitive advantage makes business attain a higher market share. The use of aluminum cans can draw consumer goodwill. The use of aluminum cans make them convenient, easy to use, durable, and light and these aspects make them a favorite aspect across many businesses.

The use of aluminum is also the aspect that it offers a better printable surface that accepts paints and coatings, and primary advantage for attractiveness and brand identification of packaging.

Governments should also come up with modes and means of trying to realize which other materials have the recycling ability linked to aluminum. Such materials should be sampled out and rules stipulated so that companies can ensure they save costs and attain more profits.

Aluminum recycling should be highly acknowledged, as it possesses the right aspects to businesses and environment. The aspect of aluminum saving more energy is an aspect that is vital to any business, as it would help reduce the costs attributed to excessive consumption and use of energy. Companies should ensure that they utilize aluminum recycling effectively.

Aluminum recycling should also be implemented as it helps save the environment. Recycling aluminum indicates that the world has been eliminated from accumulating more un-decaying materials. Companies and scientists should come up with ways of identifying how to recycle other materials that possesses danger to the environment. Aluminum recycling is also vital as it helps avoid excessive mining, which could have led to many empty holes, which would have been otherwise dangerous to the public.

Respective governments should publicize and also commercialize recycling machines to encourage many people to practice aluminum recycling as a way of helping save the environment and as a way of creating more jobs. Governments should also try and ensure that they open more avenues for depositing used aluminum cans so that companies would be able to access them and recycle them. Having more collection points for used aluminum would ensure that less fuel and costs are used and incurred by companies and that the process of recycling is shortened and that would increase the profitability of the company and increase revenues and the economy.

References

Askeland, D. R., Fulay, P. P., & Wright, W. J. (2011). *The science and engineering of materials.*

Gitlitz, J. S. (2002). *Trashed cans: The global environmental impacts of aluminum can wasting in America.* Arlington, Va: Container Recycling Institute.

Haggar, S. (2007). *Sustainable industrial design and waste management: Cradle-to-cradle for sustainable development.* Amsterdam: Elsevier Academic Press.

Inches, A., & Chambers, M. (2009). *The adventures of an aluminum can: A story about recycling.* New York: Little Simon.

MacGregor, C. (2003). *Recycling a can.* New York: Rosen Pub. Group.

Newell, J. (2009). *Essentials of modern materials science and engineering.* Hoboken, NJ: John Wiley & Sons.

Pohl, K., Acosta, T., & Gutierrez, G. (2007). *What happens at a recycling center?.* Milwaukee, WI: Weekly Reader Early Learning Libraiy.

Plastic Recycling

Abdulrahman Alresheed

Introduction

Today, the earth is perceived to be in problem. Forests are endangered, pollution is rising every day, and we are discarding the waste on our land as well as in our oceans. However, we can still save our earth by doing one thing every day to help the earth and decrease the effects of the waste. We can recycle and reuse as well as reduce the waste material in order to decrease the trash greenhouse gas emissions. Recycling has emerged as an important issue in the modern society and is amongst the issues that have gradually earned momentum over the decades as blows and dangers to the environment as well as the human health (L, 2003). With the development of technology, the quantity of substances that can be recycled or reprocessed also increased. Considering the current population trends along with the degree of wastes that will be created, it is vital that recycling and reprocessing become a daily routine activity in our lives. For this scenario, plastic is one specific material that fascinates me. A plastic is a material that is consumed more and more now days through every industry facet (Shashoua, 2012). Anywhere you look, plastics are utilized. These are utilized in packaging, constructing supplies, consumer goods and electronics apart from the hauling and adhesives (Sustainability, 2011). Considering the quantity of goods we make and expend every year and the information that a very good fraction of these products are prepared from plastics, only strengthens the requirement for better reuse plus the plastic recycling procedures. All through this research paper I will be trying to answer some common beliefs and questions concerning the recycling industry in accordance to the usage and the recycling of plastics. Further from this, I will be focusing and highlighting the uses of plastics and the chemical materials that are used in the making of plastics along with few of the disposal approaches that are used for plastics. Apart from this, I would also like to shed some light on some of the continuing arguments plus the differences over recycling plastics.

Body

Recycling- a Substantial Issue:

Considering the recycling process and the need with the significance of the subject, there are masses that are unaware of it. Therefore when the subject is brought under consideration the very first question that arises in minds is that what exactly is recycling? Why there resides a need to recycle the product? What materials and products can be recycled and what not? These are few some of the questions that are put forward by the average individual when it comes to disposing of materials to the discarded stream. It is the need of the hour to understand and perceive recycling as strong waste management strategy. This should be considered as a of solid waste management strategy that is just as useful as landfilling and incineration as well as environmentally more appropliate. "Beyond straight incineration as a waste reduction system, the British and German led the way with technology to produce energy from incineration" (Lumberg, 1989). According to Lund, recycling is today's most preferred solid waste strategy. Every production as well as the consumption activity that is involved in creating waste which is expensive to handle as well as environmentally destructive. Environmental impact of the materials is amongst one of the most important subject facing logistics along with transport managers nowadays (Sustainability, 2011). Since the previous decades an increase in the efforts to study better methods and logistics systems has been witnessed to decrease crowding, preserve natural resources as well as ease the emission. Recycling of materials like metal, paper and wood along with the plastic and glass is largely observed as an environmental friendly exercise since it protects energy, decreases raw material mining and fights climate changes. Nevertheless, few earlier studies revealed that basically plastic recycling supply chains are logistically incompetent, extensive, insubstantial, and all the more ecologically destructive. The present research paper recognizes highlights and thus maps logistical as well as the ecological courses of plastic waste.

The figure is taken from (Europe, 2009). As per the figure, the growth of the graph of energy as well recycling for the year 1996 till 2008 has escalated. It detects issues that are faced by numerous parties contributing in the collection, transportation, trading, organization, accommodation, and recycling of plastic waste, and additionally classifies opportunities and logistics keys to advance the effectiveness of plastic recycling supply chain. It includes numerous local organizations and waste companies. Making use of the cradle-to-grave tactic, also termed as the life cycle assessment (LCA) to evaluate environmental influences of post- consumer plastics during the course of their end of life cycles.

Figure shows the demands of plastics in European countries (Wong, 2009–2010). Plastics take a significant part in the environmental and societal along with the economical dimensions of ecological development (Europe, 2009). Plastics are considered light, long-lasting, sanitary and adaptable and that is the reason why they have been progressively utilized for the packaging, electronic, building, automotive and electrical goods. If any other material is utilized as to substitute plastics, the cost in addition to environmental effects is more expected to increase. For instance, Americans make use of the 100 billion plastics bags per year, that are made from approximately 12 million barrels of oil; as a substitute, the utilization of 10 billion paper bags every year indicates cut down of 14 million trees per year. The utilization of crude oil for generating plastics use up a rare resource however the use of paper signifies the reduction of the ability of the planet earth to sop up carbon dioxide (plastic2oil).

Significance of Plastics in the Modern World

As previously explained, plastics count for one of the quickest emerging groups of supplies used and thrown away in our economy these days.

Several different things are made of plastic and used in countless diverse productions. The two main businesses where plastics are used are the packing business and the building and structure business. Plastics are also needed in the transportation works, customer goods, furniture, electrical machineries and cements. For this paper packaging industry will be my main emphasis, and how that relates to reprocessing. Since plastics are major in the consumer/packaging industry they will have consequences on recycling charges. The rise in plastics may result in grave problems for left-over removal processes, which are mainly managed by native administrations (Melosi, 2005). As landfill space shrinks and new landfills become nearly unbearable to spot, there is a need for the solid-waste planners should be taking in account other approaches for the waste management. The usage of plastics in wrapping has gradually amplified over time with forecasts climbing well above 20 billion pounds per year. Statistics for manufacture and construction resources are a little lesser since the use of plastics in this torrent has a prolonged lifetime, usually, 25 to 50 years. Buyer and organized goods make up the third largest group of plastics. This group comprises all takeaway agreements, removal packages used in hospitals, pens prisons; throw away razors, watches camera and lighters. By volume, it is assessed that wrapping resources version for more than one-third of community dense left-over in the United States. The wrapper business is the principal customer of plastics using one-third of all dammar produced yearly for movies, soft beverage ampules, physical on thermoses and inflexible vessels and for coat on many other protective materials. The two chief groups of current use are packing movie and rigid vessels, account for 35 percent and 51 out of a hundred of plastics, respectively. The further two groups behind these are coatings which are itemized at 9% and closures itemized at 5%. Plastic packet materials are collected of a diversity of different gum and resin blend of which the most mutual are polyvinyl chloride (PVC) polypropylene (PP), low-density polyethylene (LDPE) as well as the high-density polyethylene (**HDPE)** and polystyrene (PS) and polyethylene terephthalate (PET). Polyethylenes, high and low texture, make up over 60 out of a hundred of the plastic wrapping left-over river. The low density polyethylene film is utilized in applications for instance the grocery bags as well as the bread wraps, whereas the high density polyethylene is utilized for over 50% of plastic containers used for the milk and water along with bleach and laundry detergent. Polyethylene terephthalate (PET), mainly used **in** soft-drink flasks, establishes about 14 in each hundreds of plastic flasks and is now the most commonly used soft-drink containers, based on the total size of soft beverages sold in the United States. The quantity of resins used in these soft-drink flasks doubled over by 1995 with estimations close to 2 billion pounds being used. This number is predictable to progressively increase as plastics are used more in the packaging business.

Other resources used in wrapping, to a smaller degree, include polyvinyl acetate like the adhesive and ethylene vinyl acetate (EVA) copolymer ethylene vinyl alcohol (EVOH) as an oxygen blockade. The usage of these resources reproduces a significant recent growth in plastic wrapping - the use of compound, or multilayer packaging. These bundles contain coatings - sometimes as much as 12 - of different types of gamboge and other materials. For instance, a squeezable ketchup bottle contains another adhesive layer along with another PP polypropylene **(PP)** layer, an adhesive layer, a layer of EVOH which is the oxygen barrier layer. The usage of the EVOH lets the manufacturers to bundle in plastic many nourishments that they formerly couldn't, for the reason that of possible pollution. Many compound packages are now substituting materials with high recycling histories, such as glass and paperboard. The details for these changes include amplified client accessibility, lengthier shelf life and lighter load. However, the disposability of the parcels and the ecological impressions of removal have been largely unnoticed. Merely 5 recycled and reprocessed bottles of the soft drinks have the ability to make fiberfill enough for a ski jacket. Approximately 1,050 reprocessed milk pitchers can be whole into a six-foot bench for a park. "There is also potential danger to marine ecosystems from the accumulation of plastic debris on the sea floor" (Derraik, 2002). In United States, sufficient plastic film is made every year that it can shrink-wrap the Texas state. Only if 10% of Americans accepted products with fewer plastic packaging only 10% of the time, about 144 million pounds of plastic could be removed from our landfills.

Identifying the Need of Plastic Recycling

Identifying the significance of plastics with the fact that plastics are composed of the limited resources, a huge amount of efforts are put in the research and development for making the plastics reusable as well as recyclable (Anthony L. Andrady, June 2009). In accordance to the Department for Environment, Food and Rural Affairs (Defra)'s Waste Strategy 2007, by 2015, the government of United Kingdom has headed to accomplish 45 percent recycling target. From the peliod of 2008 to 2009, 27.3 million tons of public waste was gathered by UK local authorities however 50.3 percent of the total was sent to landfill, 36.9 percent was reprocessed or composted and only 12.2 percent was burned for energy recovery. Even with the awareness of the fact that plastics are challenging to be degraded naturally and that is the reason why United Kingdom is tossing away four out of every five plastic bottles (RECOUP, 2003). There isn't much emphasis and resides a lack of research on the end of life (EOL) management of products that are made of plastics and various other scarce resources.

Numerous researches have been done in many distinct disciplines endeavoring to uncover the technologies and other techniques to make the world a cleaner, better and sustainable world. Considering a simple question for instance the use of plastic or paper bag for the super market shopping, leading to the further complicated questions concerning the most justifiable tactics to design, construct, dispense, and reutilize a product, additional research is needed to help logistics and supply chain managers for making the updated decisions. The problem with this approach is that the major part of the research efforts is done in isolation not considering a cradle-to-grave and the life-cycle tactic. However recycling is considered to preserve materials as well as decrease the greenhouse gas (GHG) release, reutilizing activities encompass transportation along with the production activities which take in energy and natural resources thus simultaneously producing emissions and pollutions. Lacking the complete awareness of the environmental influences of recycling logistics systems, the managers won't be capable of making better decisions on product design, production, distribution, choice of materials, and the design of recycling logistics systems. Completely comprehending the environmental influence of numerous logistics solutions for dealing with the product life cycle containing the product end of-life (EOL) is a critical step in the direction of a cleaner as well as sustainable world. At present, in our economy, plastics amount to one of the rapidly growing classes of materials utilized as well as thrown away. Considering the total volume of municipal solid waste stream, plastics make up approximately 10 percent of the total weight whereas around 33 percent of the volume. In accordance to the Environmental Protection Agency (EPA), in United States there are roughly 40 billion pounds of plastics are produced. Even though considering the weight, plastic contributes merely 4 percent of the total municipal solid waste however it comprises 12% of all the packaging. The basic problem for the reusing the post-consumer plastic is the huge cost incurred for the transportation as well as collection. Mainly the empty plastic packages as well as the containers use a large volume and therefore small towns are not able to gather enough tonnage to endure their reusability as portion of an economically practicable program. According to the statistics, Americans every hour throw away around 2.5 million plastic bottles and comparatively only a very small percentage of these bottles are recycled. One of the proposed solution demands the plastics manufacturers to help sections in financing choppers and extra equipment that decrease plastics into a thicker form that is further cost-effective to transport (Lund, 2001). In the same way, with assistance of the American Plastics Council (APC), the plastics industry has funded research intended at cutting down the expenses of collecting, handling, sorting and recovering plastics. Nevertheless, regardless of the costs, recycling endures to be highly appreciated by the American public as well as the communities carry on to extend collection services for recyclables (APME, 2004).

Plastic recycling requires five basic parts for as to be effective. The five parts are known as the collection, processing, reclaimers, end users and the last being the customers. Collection packages are required for businesses, industries and the overall public. For the processing part, processors locally formulate collected plastics for the marketplace by densifying as well as sorting it into a cost-effective form for shipment to the long-distances. Reclaimers convert the convalesced outcome into feedstock materials. Then it is the job for the end users to transform the reclaimed material into reprocessed products and therefore in the end it is the job of the consumers to purchase the recycled substance products. The arrangement and the infrastructure for these five steps i.e. the reclaiming, processing and the collection of the post-consumer plastics raised immensely starting from the late 1980s till the late 1990s. Prior to 1998, recycling of the ·post-consumer plastics involved predominantly soft drink bottles gathered from a few states demanding deposits, and the plastic covering from resumed lead-acid automobile batteries (Lund, 200 I).At the moment, there are several community collection platforms that propose pick-up as well as the drop-off spots for exclusive cate-g0lies of plastic products for 80% of the United States. On the other hand, repossession of post-consumer plastics because of the community collection programs has never kept speed with reclamation capability and end-user requirement for post-consumer plastics. Consequently, these plants function well below their capability.

The main recycling of plastics, which preserves the greatest quantity of energy, consists of the reconversion of sanitized plastic waste to its initial pellet or mastic form. The recycled form contains chemical as well as the physical characteristics identical to the characteristics of the original product. Primary recycling is appropriate essentially for industrial wastes, as they generally comprise of one kind of resin and are not usually polluted by use in consumer precuts (Wolf, 1991). Secondary recycling transforms scrap plastic or also the waste into goods possessing characteristics which are far less challenging in comparison to the original product. This procedure makes use of the plastic wastes, generally post-consumer wastes which is not considered appropriate for direct reprocessing because of the contamination level that is present. The third type called the tertiary recycling typically transforms the plastic waste into a fuel intended for chemicals as well as for direct energy and last type of cycling that is the quaternary recycling is the actually is a process of generating energy from the flaming of plastic wastes (Initiative).

As discussed above in the research, the main point regarding the recycling post-consumer waste plastics is that, not like materials for example glass and metals, plastics could not be decontaminated and purified at present and recycled for food packaging. Therefore, for instance, reprocessed plastic soda bottles cannot come back in the form of other bottles for the sodas or any other food items but as pillows or fiber or plastic lumber, etc. Even though it is looked-for to recycle the maximum plastic as possible still the case is that most plastics can never be utilized in the closed-loop recycling so beneficial to other post-consumer supplies (M., 2003). Even though much development has been made still the recycling of post-consumer plastics is restricted by the nonexistence of a strong recycling arrangement. The leading obstacle to improved recycling include: the deprivation of economic feasibility collection, segregation, and transportation procedures; the lack of large-scale marketable recycling operations proficient of tackling a heterogeneous blend of contaminated post-consumer supplies; and the absence of steady and challenging markets for the recycled plastic products (Wolf, 1991).

In United States, the Solid waste policy is directed at developing as well as implementing proper procedures to efficiently tackle the solid waste. In order to make the solid waste policy an effective policy; the stakeholders as well as the citizens, and government agencies along with other research organizations should give in their input. EPA- The Environmental Protection Agency of United States of America aims at safeguarding the environment along with the human health by ensuing laws passed by congress that help sustain a healthy environment. The policy covers the waste facility licensing, management planning and how to pack as well as manage the hazardous waste that is produced (EPA). The policy requires the waste management and the industries to manage their waste so that they do not pollute the environment by polluting the water as well as the air. Proper check on the industries and factories is maintained for ensuring the healthy environment.

Further from this, the pyrolysis is the procedure deployed for the conversion of scrap plastic to other fuel sources. Twenty-three manufacturers were identified of the PTF technology. Every technology has a unique perspective with regard to the kind of scrap plastics which the systems can tackle, along with the output or else the fuel product. These systems that help in the conversion of the scrap plastic to the fuel sources have some common features. These include certain level of pretreatment, which could be as less involved as size reduction or in cases as engrossed as cleaning along with the moisture removal. Second is the conversion. For this this purpose the pyrolytic processes are utilized in order to transform the plastic into gas. Next comes the distillation in which the gas is converted to the liquid form (Hopewell J., 2009).

Plastic Resource Recovery

Resource recovery already has and in the future will become more common as well as demanding as landfill room is finished. The waste alternative for plastics, substitute to landfilling exists as modern incineration or else can be called resource recovery (Denison, 1990). Incineration recuperates energy by means of combustion, which is transformed into steam or electric power. Resource recovery plants are not different or unfamiliar and have been in process for few years now in the United States and Europe along with Japan. Nevertheless there are countless environmental threats allied to these plants for example gas stack emissions or landfilled bottom ash holding toxics or heavy metals excess after the incineration process. These environmental threats have increased enormously owing to the fact that many plastics are fabricated of a number of diverse chemicals for instance plastic resins, shields, plasticizers besides additives. Even though according to the general view, plastic waste has an extraordinary energy content, also burns powerfully, resulting in petite residue, there still reside countless unanswered questions regarding the consequences of burning and what these influences have on human health as well as the environment (Jefferson Hopewell, 15 June 2009). Generally incinerators are mass burn incinerators in which 85% of the total volume of wastes is wiped out.

Yet, emissions that are resulted from these mass bum incinerators may as well contain acid gases, particulates, oxides of nitrogen, heavy metals as well as trace organic compounds.

Conclusion

To conclude, in my opinion I think it's obvious that recycling is very significant and an issue that every family, community, business, etc. should hold. Mainly with the plastics industry keeping in mind the quantity of plastics this country consumes and produces every year. This is appreciable and amazing as well just to what extent our society has transformed with the initiation of plastics for industries for example manufacturing, construction and building, transp01iation in addition to most essentially, consumer goods. A change has occurred in us as a nation to exist like a disposal civilization, a society that explores something unpretentious and easy. Education also plays an important part in the recycling process of plastics as well as other materials that are a potential hazard to the society. On one hand while there are products that are reaching the shelves in highest numbers packed or comprising plastics, it's crucial for us to recycle as well as to endure to develop technologies as to fight the excessive wastes we produce. I am of the perspective with the research in my mind that pollution prevention is the significant factor to support alleviating the pressures of waste when it approaches materials that are required to be, or should be recycled, as well as handled carefully.

Bibliography

Anthony L. Andrady, M. A. (June 2009). Applications and societal benefits of plastics. *Phil. Trans. R. Soc. B.*

APME. (2004). *An analysis of plastics consumption and recovery in Europe.* Brussels: Belgium: Association of Plastic Manufacturers Europe.

Denison, R. A. (1990). *Recycling & Incineration: Evaluating the Choices.* Washington, D.C.: Island Press.

Derraik, J. G. (2002). *Thepollution of the marine environment by plastic debris: a review.* Marine Pollution Bulletin 44.

EPA. (n.d.). *EPA.GOV.* Retrieved from United States Environmental Protection Agency: www.epa.gov

Europe, P. (2009). *The Compelling Facts About Plastics 2009.*

Hopewell J., K. E. (2009). Plastics recycling: challenges and opportunities. *The Royal Society.*

Initiative, G. M. (n.d.). *Plastic to Fuel Summary.* Retrieved from http://wmich.edu/mfe/mrc/greenmanufacturing/pdf/PTF%20 Summary%20Updated.pdf

Jefferson Hopewell, R. D. (15 June 2009). Plastics recycling: challenges and opportunities. *The Royal Society(364).*

L, A. A. (2003). *Plastics and the environment.* West Sussex, England: John Wiley and Sons.

Iumberg, L. a. (1989). *War on Wastes: Can America Win its Battle with Garbage? .* Washington, D.C.: Island Press.

Lund, H. F. (2001). *The McGraw-Hill Recycling Handbook: Second Edition.* New York: McGraw Hill.

M., F. (2003). Plastics recycling. In *Plastics and the environment* (pp. 563-627). NJ: Wiley Interscience.

Melosi, M. V. (2005). *Garbage In The Cities: Refuse Reform and the Environment.* University of Pittsburgh Press.

plastic2oil. (n.d.). *Converting Waste Plastic to Ultra-Clean, Ultra-Low Sulphur Fuel.* Retrieved from plastic2oil: http://www.plastic2oil .com/site/home

RECOU P. (2003). *Plastic economy in the UK - a guide to pol ymer use and the opportunities for recycling.* RECOU P.

Shashoua, Y. (2012). *Conservation of Plastics.* Routledge.

Sustaina bility, 4. (2011). *Conversion technology: A complement to plastic recycling.*

Wolf, N. a. (1991). *Plastics: America's Packaging Dilemma.* Washington, D.C: Island Press.

Wong, D. C. (2009–2010). *A Study of Plastic Recycling Supply Chain.*

Plastic Foam Recycling

Ahmed Baroom

Introduction

Foamed plastic is formed through the process of conversion of synthetic resin into a sponge-like substance that has structures either closed or open celled, and which can be either rigid or flexible. Foamed plastics are utilized in making numerous products including materials used for cushioning, air filters, furniture, toys, thermal insulation, sponges, plastic boats, building panels, and many more things. There are numerous plastics that are commonly foamed and these include vinyl, polystyrene, polyethylene, phenolics, silicones, and cellulose acetates among many others. The production of foam plastics composed of closed-cell structures is through incorporation of a blowing agent that, with decomposition at the plastic's fusion point, releases gas bubbles that are trapped during the process of gelling (Avella, Cocca, Errico, & Gentile, 2011). The production of plastic foams with open-cell structures is through the incorporation of an inert gas into the resin under pressure, and then the mixture is released to the atmosphere and the resulting foam is cured.

Discussion

The recent trends by cities across America, notably in contemporary New York, of banning the city's establishments from using plastic foam containers because of the harm that they cause to the environment endorse the significance of recycling. The materials have been condemned as environmental hazards that clog the landfills of America. The ban of these materials is hoped to significantly improve the rivers and waterfronts of America and thus will enhance the marine lives as well.

It is acknowledged that the best means to limit the plastic wastes that businesses and homes release and to prevent such wastes from going to landfills is to avoid their use entirely, or engage in reusing and recycling. Far from eliminating, the use of plastics in everyday life is actually increasingly becoming the norm. Thus recycling and reusing are is considered to be substantial strategies. Recycling of foam plastics requires that we should engage in identification of the types of plastics because varying types of plastic undergoes different methods of recycling. Common foam plastic products include polyethylene terephthalate, which is mostly used to package soft drinks and fruit juices. High density polyrthylene are plastics used in packaging as milk bottles and shampoo containers. Polyvinyl chloride or plasticized polyvinyl chloride are used in cordials, juices and squeeze bottles. Low density polyethylene is mostly used as garbage bags and bins (Chanda, & Roy, 2009). Polypropylene is commonly used as ice cream containers, and is used to package takeaway food items as well as lunch boxes. Polystyrene is commonly used as yoghurt containers, as plastic cutlery and foamed hot beverage cups. Plastic bags, foamed plastics lining bins and polystyrene foam plastics are not recyclable.

Advantages of Plastics

There are numerous advantages that are offered by use of plastics. Plastics are light and robust and can therefore be used in production of products that though strong, are also light, unlike other materials such as ceramics and metal. Plastics also have a resistance to rust and corrosion. As they are resistant to acids, alkalis, and oils and therefore corrosion and rusting does not take place. Plastics also have the appeal of being mass producible. This property allows plastics to be molded and processed in numerous different shapes so that there can be efficient mass production of products with complex shapes. Mass production of plastics has helped to drastically decrease the cost of the products. Plastics also possess excellent insulation properties that increase their preference in electronic and electrical products (Chang, et al., 2013). They also provide high efficiency in insulating materials from heat. Plastics are also preferred by restaurateurs because of their cleanliness and the fact that they are impermeable to oxygen, making them very effective in protection of foods from contamination by microorganisms.

Impact on the Environment

Plastics have various impacts on the environment. They emit toxic greenhouse gases because of the fossil fuels that are used in the production of plastics. The decomposition of gases also leads to production of greenhouse gases, particularly methane. Methane is considered to be 20 times in strength as compared to carbon dioxide and represents a significant part of emissions from landfills. Since plastics are made from natural resources, their depletion means that they cannot be replaced. Plastic is very persistent in the environment and most plastics are not biodegradable and thus they will survive in the environment for many decades. Despite best attempts at recycling, a significant portion of used plastics still end up in landfills leading to adverse environmental impacts (Niaounakis, 2013). Plastics also present about 80% of wastes that are dumped in the global oceans. Plastic wastes are a serious threat to marine life. They cause injuries and even death as they are mistaken to be foods by marine life. Estimations place the impact of marine rubbish to be the death of over 100, 000 mammals annually and more than a million seabirds.

There has been an increase in effective utilization of plastics every year due to widespread utilization of mechanical and feedstock recycling as well as increased generation of solid fuel from wastes, efforts made at waste power generation and increased development of heat-utilizing incinerators. It has also been observed that there has been increased utilization of significant quantity of plastics in mechanical recycling. Mechanical recycling refers to the use of plastic materials as raw materials in the production of new plastic materials. Mechanical recycling is widely utilized because a substantial portion of industrial waste plastic is appropriate for this kind of recycling, due to their quality and because of stable supply of such plastics (Saddleback Educational Publishing, 2009). The increase in mechanical recycling is also attributed to the sustained transition to containers, packaging materials, and domestic appliances that are recyclable.

Methods of Plastic Recycling

Material recycling is the use of plastics as raw materials in production of new plastic products. Chemical recycling is made possible by such methods as monomerization, blast furnace reducing agent and chemical feedstock recycling by using a coke oven. Thermal recycling utilizes cement kiln, and is highly utilized in waste power generation. The purpose of recycling is to curb the consumption of limited natural resources and mitigate impact on environment. The selection of a recycling method that has the least social costs and effectively limits adverse environmental impacts is highly recommended.

Mechanical Recycling

Mechanical recycling is used for plastic recycling in industrial contexts as the generation of industrial plastic waste in the processes of manufacturing, processing and distribution of plastic products provides plastics as ideal raw materials due to distinct indication of the type of resins, less dirt and impurities, and their availability in large quantities is guaranteed. Plastic products after collection are sorted for elimination of impurities and then they are shredded and cleaned. There is also the removal of foreign bodies and materials that are non-resins. The products are then made into flakes and pellets. The recycled materials are then taken to textiles and plants where sheets are made so that they can once again be melted in production of textile and sheet products.

Mechanical plastic recycling also takes place through resin moulding techniques. Extrusion molding refers to the process where there is melting of resin and its continuous extrusion through a mold using a screw to produce a molded product such as pipes, films, wire coverings etc. Injection molding is where there is injection of resin, taken through heating, melting, and solidification resulting into formation of molded products such as plastic buckets, washbowls and bumpers. There is also blow molding, which refers to obtaining a parison through clamping of extrusion or injection molding into a mold and inflation with air for production of all bottle models (Gent, Menendez, Toraño, & Diego, 2009). Plastic bottles are made using stretch blow molding in order to make them less likely to rupture. Vacuum molding refers to where a sheet softened through heating is sandwiched in a mold and there is sealing of the space between the sheet and mold and evacuated for production of materials such as cups and trays.

Monomerization

Polybutylene terephthalate (PET) plastic bottles cannot be used in making drinking bottles. This is because they are unsuitable as raw materials for soft drink, alcohol and other types of bottles because of hygiene and smell concerns. Conversion of PE bottles into their previous forms is considered to be more economical than production of new products from petroleum and naptha. Monomerization causes chemical decomposition of used PET bottles into their constituent monomers through depolymerization. The components are molded into new PET bottles. Another method developed by Tenjin Limited entails the combination of ethylene glycol and methanol for breaking the waste resin found in PET plastics into dimethyl terephthalate for conversion into the raw material that is utilized in production of textiles and films (Gonella, Zattera, Zeni, Oliveira, & Canto, 2009). There is also the method of manufacturing resin by its breakdown into highly pure bis hydroxethyl terephthalate monomer through a novel technique of depolymerization, by utilizing a method known as EG developed by Aies Co., limited.

Blast Furnace Feedstock Recycling System

In this process, plastics are utilized as a reducing agent. Collection of plastic wastes from factories and homes is undertaken and they are cleansed of non-combustible matters as well as other impurities such as metals. The plastic wastes are then finely pulverized and they are packaged for reduction of their volume. There is granulation of the plastics that do not contain PVC and they are then fed into the blast furnace after the separation of hydrogen chloride at very high temperatures of about 350 degrees Celsius in the absence of oxygen, as emission of hydrogen chloride will have massive impact on the furnace (Sendijarevic, 2007). The extracted hydrogen chloride is recovered as hydrochloric acid and utilized for different purposes. This is the dehydrochlorination method developed by the Japanese Plastic Waste Management Institute.

Another method of recycling involves the use of plastic wastes as chemical feedstock and fuel. This method would involve collection of plastic wastes which are then made to undergo shredding and removal of impurities such as iron. PVC removal is done and then granulation of the plastics and their mixing with coal takes place. The end product is fed into the carbonization chamber of a coke oven. Combustion of the plastic does not take place inside the carbonization chamber due to absence of oxygen (Niaounakis, 2013). The plastic inside the chamber is rather cracked at very high temperatures to produce coke, which is utilized as a reducing agent in coke ovens.

Gasification is another chemical recycling method. The gasification process entails heating plastics and then addition of a supply of oxygen and steam. There should be limited supply of oxygen thus implying that most of the plastic will be transformed into hydrocarbon, carbon monoxide, and water. Liquefaction is another chemical recycling method. Liquefaction is a process that converts plastic wastes back to oil as plastics are produced from petroleum (Zhao, Zhao, & Chen, 2015). Plastic recycling cannot be effective without elaborate collection and sorting mechanisms.

Barriers to Plastic Recycling

Domestic plastic waste collection includes curbside collection, use of designated drop-off locations—among others. It has been observed that waste collection schemes that require residents to take wastes to deposit sites lead to lower collection rates with exception of instances where there is high public commitment or there is a direct economic incentive. Changing consumer perceptions and attitudes towards recycling is required, but they also need to be guided on the types of plastic packaging that are recyclable. Research has shown that there is demand by residents for collection of various plastic types including carrier bags and food packaging, among others.

It has also been acknowledged that the barriers to recycling are majorly related to information distribution and the efforts required toward rinsing out residues from containers first. They also need to be provided with evidence regarding the impact that their actions are making. Improvements in the collection and sorting of waste plastic materials will also lead to efficiency in recycling. This is because one of the dominant challenges facing plastic recycling is that plastic is immiscible at the molecular level and thus it demands distinct processing requirements. This therefore requires that plastics should be sorted into their single types so that there can be cleaner and fewer types of plastics, which can ease the challenge of recycling.

Bio-based plastics are generally not considered to be a challenge in mechanical recycling as their design allows them to keep their properties. Biodegradable plastics, in comparison, lose their characteristics creating a lower melting flow index that refers to the rate of decomposition of organic components. Alternatives to recycling include incineration of plastic wastes and disposal at landfills. Energy recovery from plastic waste materials should be on the products that are non-recyclable. Technological advances have also led drastic reductions in cost of recycling by enhancing efficiency and have been instrumental in facilitating closure of the gap between the value of recycled and virgin plastics. For instance, a way through which the value of recycled plastic can be enhanced would be through technologies that transform recovered plastic into polymers that are of food-grade, through elimination of contamination. Such a technology has been proven for clear PET bottles.

There are a number of barriers to the recycling of plastics. These include constraints due to lack of guarantee of supply—plastics being vital components of certain existing or new companies—and constraints in financing for commercial development. Financial and technical barriers place a limitation on the ability of the smaller recyclers to accept a wider range of the foam plastic types that are emerging or conversion into other products that are more consumer-friendly. The end result is that there are variations in recycling programs and capabilities that leave consumers with confusion regarding the materials that are recyclable and those that are not.

Growth Trend

Plastic recycling is a sector that is poised for growth with increased investment and acceleration in market development. There are numerous factors that justify the need for immediate investment in this sector. These factors include aggressive demand for recycled plastics, the continued development of technologies that would enhance the rates for recycling plastic materials and advances in technologies used in processing foam plastic materials, which have created a platform for the cost-effective recycling of various types of foam plastic materials. The capacity of further growth for most manufacturers is inhibited by the lack of a secure and steady source for supply of recyclable plastic materials that are clean and have been sorted. Foreign export markets are expected to reduce in their significance thus creating a gap and an opportunity for local markets and consequent enlarging of local processing capacity (Kutz, 2011). Plastic recycling is a potential source for economic contribution that has not been fully tapped and can contribute significantly to job development.

It is also acknowledged that the alternatives to recycling, such as conversion of waste plastics and conversion of plastics to oil have the potential to utilize significant quantities of foam plastic materials that are non-recyclable either due to over contamination or that are non-recyclable. Recovery of such non-recyclable waste foam plastic materials and their productive utilization has the potential to create opportunities for economic development and entrepreneurship. There are numerous suggestions that are advanced on the manner in which supply of recycling materials can be attained. These include developing partnership with members of the private sector, promotion of waste exchanges and setting up market places for plastics. Exhibitions for products made out of recycled plastics would also help to show people the potential of recycling plastics.

Conclusion

This paper has explored the various approaches in fronting the topic of foam plastic recycling. Plastic is increasingly becoming the dominant material of choice of product designers. The variety of different types of plastics is critical and should be taken into account when planning waste collection, sorting, and processing activities. This is because plastic polymers are immiscible at the molecular level and thus contamination makes the task of recycling difficult and challenging. It has been identified that one of the major challenges to recycling is poor knowledge of potential for plastic recycling at either the regional or local level. The advances in technology have also been noted to herald new opportunities to recycle wastes that previously would have been thought of as being non-recyclable.

Technological developments in sorting and processing of waste plastic materials would also increase the rate and efficiency of recycling. Recycling waste materials is also an imperative in order to reduce wastage of natural resources that are used in their production, to effect significant energy savings, enhance the health of marine water bodies and marine life, as well as limiting the emissions of greenhouse gases that are released in landfills. Waste plastic recycling requires the awareness of the broad range of methods that are also available, ranging from mechanical and chemical to thermal methods of recycling plastics or using them as materials for energy production.

References

Avella, M., Cocca, M., Errico, M. E., & Gentile, G. (January 01, 2011). Biodegradable PVOH-based foams for packaging applications. *Journal of Cellular Plastics, 47*(3), 271–281.

Chanda, M., & Roy, S. K. (2009). Plastics fabrication and recycling. Boca Raton, FL: CRC Press.

Chang, Y. M., Liu, C. C., Dai, W. C., Hu, A., Tseng, C. H., & Chou, C. M. (2013). Municipal solid waste management for total resource recycling: A case study on Haulien County in Taiwan. *Waste Management & Research: the Journal of the International Solid Wastes and Public Cleansing Association, Iswa, 31*(1), 87–97.

Gent, M. R., Menendez, M., Toraño, J., & Diego, I. (2009). Recycling of plastic waste by density separation: Prospects for optimization. *Waste Management & Research: the Journal of the International Solid Wastes and Public Cleansing Association, Iswa, 27*(2), 175–87.

Gonella, L. B., Zattera, A. J., Zeni, M., Oliveira, R. V. B., & Canto, L. B. (2009). New reclaiming process of thermoset polyurethane foam and blending with polyamide-12 and thermoplastic polyurethane. *Journal of Elastomers and Plastics, 41*(4), 303–322.

Kutz, M. (2011). *Applied plastics engineering handbook: Processing and materials.* Amsterdam: William Andrew.

Niaounakis, M. (2013). *Biopolymers: Reuse, recycling, and disposal.* Amsterdam: Elsevier Science.

Saddleback Educational Publishing. (2009). *Recycling.* Irvine, CA: Saddleback Educational Publishing.

Sendijarevic, V. (2007). Chemical recycling of mixed polyurethane foam stream recovered from shredder residue into polyurethane polyols. *Journal of Cellular Plastics, 43*(1), 31–46.

Zhao, Q., Zhao, Q., & Chen, M. (2015). Automotive plastic parts design, recycling, research, and development in China. *Journal of Thermoplastic Composite Materials, 28*(1), 142–157.

QUESTIONS ABOUT RECYCLING

1. What things have you learned about recycling from reading this chapter?
2. What things do you recycle now?
3. Talk about what your home community does to support recycling.
4. Discuss changes you believe should be made to make recycling more popular.

Chapter 8

Pollution

This chapter continues the study of environmental issues by placing the lens on pollution. Three authors have written about some sources of pollution that our society is currently dealing with. Solid waste stream pollution deals with most things that we as consumers discard, as well as the solid wastes from manufacturing and construction. What we do with things when we no longer need or want them plays a big role in whether they are a significant source of pollution of our air, water, and soil.

In the article on landfill design, you will learn about how landfills are designed to contain pollution as securely as possible, to protect our air, land and drinking water.

The third article discusses one specific industry, and the pollution it produces. Paper mills have long been a source of unpleasant smells that prompt residents in neighboring areas to complain. To find out whether those foul smells are really harmful, please read the article.

Solid Waste Stream Pollution

Musaad Alrawili

Solid waste is a term that is used to refer to everyday substances we use, consume and then throw away. Such items are no longer useful and include wasted food, polyethylene bags, electronic accessories, many forms of plastics, old tattered clothes—among many others. Also, parts of liquid waste materials are also considered as solid wastes. Solid waste stream pollution is when the streams and rivers are filled with non-biodegradable and non-compostable biodegradable wastes that have negative effects on water bodies. Although there are laws and regulations that safeguard the quality of streams and rivers, solid waste in the form of plastics, sediments, garbage, and trash in most cases ends up in these streams and rivers. Generally water collects in depressions and hence solid waste that is dropped or blown into such a waterbody will eventually reach a stream or a river in form of a drainage way. Trash and litter (general terms for dry solid waste) in urban areas are transported by surface water runoff. In both urban and rural areas, solid waste is sometimes illegally disposed directly into streams or rivers. Trash can also be from fishermen or from people who participate in water recreation activities like swimming or boat riding. Regardless of the type of solid waste, trash is a form of water pollution and is harmful to living organisms.

Ironically, some form of discarded solid wastes like plastic containers, tires, and non-organic construction debris provide suitable habitat for some aquatic organisms. On the other hand, trash items are an eye sore and suggest human negligence and disrespect for aesthetic values of the environment and natural ecosystems. Despite increased environmental awareness by the government through the mass media and public campaigns, some people still use rivers and streams as a depository for solid waste, including couches and mattresses, electronic equipment, medical equipment, clothes and plastic containers.

However, when solid wastes accumulate in streams and rivers they disrupt the ecosystem of the marine life, such as the fish, turtles, and even marine plants.

Classification of Solid Wastes

- **Biodegradable waste:** these include left overs of food and kitchen waste, plant residue waste, papers like newspapers.
- **Recyclable material:** these include polyethylene paper, glass, plastic bottles and cans, metals, fabrics, clothes, batteries etc.
- **Inert waste:** these include waste derived from buildings and constructions and also demolition waste, dirt, rocks, debris.
- **Electrical and electronic waste:** electrical appliances like sockets, extensions and equipment like television sets, computers, and monitors would be some examples.
- **Composite wastes**: these include waste clothing from textile industries, tetra packs, waste plastics. such as toys.
- **Hazardous waste:** these include nearly all forms of paints, petrol chemicals, light bulbs and fluorescent tubes, containers, fertilizers, spray cans.
- **Toxic waste:** these include harmful chemicals that are used by human beings in daily activities like nematicides, pesticides, acaricides, fungicides, herbicides.
- **Medical waste:** these are wastes from hospitals, which are disposed in the landfills, e.g. used syringes, used cotton wools and bandages, expired medicine, scalpels etc.

Repercussions of Solid Waste on Streams
Death of Aquatic Organisms

The main problem caused by water pollution is that it kills life that depends on these streams and rivers. Among the organisms that depend on streams and rivers are fish, crabs, birds, insects and others. When organic wastes are produced by people and animals, they include things like fecal matter, yard clippings, crop debris, plastic cans, food wastes, rubber, and wood among others. In presence of oxygen such solid wastes will decompose. When they are discarded into rivers and streams and begin to decompose, they will definitely deprive aquatic organisms of the oxygen the latter require for their survival hence leading to their death. Although there is no expert consensus about the impact of solid waste on surface and underground water sources, some experts argue that even common solid waste items like polyethylene bags and newspapers pose a considerable risk to the quality of water. Other experts still argue that the impacts of landfills on underground water would be insignificant if perilous materials for example paint, motor oil, chemicals and incinerator ashes were banned from the underground water sites (Johnson, 1978 and DeLong, 1993).

Affect the Reproductive Behavior of Marine Life

Organic waste (e.g., wood wastes) can have chemical and biological effects on rivers and streams. For instance, they may interfere with the establishment of aquatic plants, which in turn affects the reproductive behavior of fish and other animals, and also deplete the water of dissolved oxygen as the wastes decompose. This in the long run interferes with the spawning of fish, hence interrupting the entire cycle of fish reproduction.

Inhibit Penetration of Light

Activities like bush clearing, mining, grading, excavating, and filling of land make the soil to loosen, making it prone to agents of erosion. When there is heavy rainfall, it causes erosion to occur on the land thus making sediments to be deposited into rivers and streams. According to Philip R. O'Leary, and Patrick H. Walsh (1995), various mining activities can lead to minerals such as zinc, tin, chromium, platinum, iron, copper, etc., to be discharged into streams and rivers. When excessive levels of sediments and minerals are transported and deposited in water they inhibit the penetration of sunlight, which in turn reduces the production of photosynthetic organisms. This will in the long run hinder their growth and development.

176

Lowers Quality of Water

Some solid wastes like polyethylene papers, newspapers, paint cans—when thrown in rivers, lower the quality of water. This is because they contain toxic chemicals and other perilous materials. The environmental protection agency (EPA) has recognized approximately 582 extremely poisonous chemicals, which are produced, manufactured, and stored in locations across the United States. Some chemical manufacturing industries incinerate toxic waste, which in turn produces hazardous by-products like furans and chlorinated dioxins, which are extremely dangerous (Rogers, William, 1986). When such toxins mix with water they can form other deadly compounds which when they find their way into rivers and streams are very hazardous. Other solid wastes that are produced or stored by households include used motor oil, among others. However, some solid wastes when disposed in water tend to react with it. For instance, some can dissolve in water and make the rivers water to discolor. This will have adverse effects because for one, discolored water destroys the aesthetic nature of the environment. Also, such water can be highly toxic to an extent of being of no economic benefit to the human beings. This is high level of destruction of natural resources.

Effect on Human Health

Some solid waste disposed in streams and rivers have poisonous substances that can have negative impact on human health, with children being more vulnerable to these contaminants. Habitual drinking of water that is contaminated leads to communicable and chronic diseases like diarrhea and typhoid. This is because release of chemical waste into the streams makes it prone to disease causing pathogens (Gregory S. Weber, 1997). Organic domestic waste disposed in any uncontrolled manner poses a serious threat to the health of people. The major cause for this is that the biodegradable materials decompose and ferment under uncontrolled and unhygienic conditions thus creating conditions that are favorable to the survival and growth of microbial pathogens. According to a World Health Organization report, diarrhea caused by this type of contaminated water leads to millions of deaths each year worldwide.

Disruption of Food Chains

Pollution disrupts the natural food chain as well. Some solid wastes can contain minerals such as lead and cadmium that are eaten by microscopic animals. Later, these animals are consumed by some aquatic organisms like fish. Human beings will consume such fish that is highly toxic hence inducing that poison in their blood stream. The worrying thing is that in some cases consumption of some organisms can lead to death of other organisms. This will lead to increase of particular organism because there are no predators. Therefore the food chain continues to be disrupted at all higher levels if the rivers and streams continue to be polluted by solid wastes.

Can Lead to Desertification

Some solid wastes contain nutrients like phosphorus and nitrogen and support the growth of algae and other plants forming the lower levels of the food chain. However, excessive levels of nutrients from solid wastes with hormonal chemicals can lead to eutrophication, which is the overgrowth of aquatic vegetation. For instance, this vegetation spreads over water and weakens some plants. Over time, this overgrowth of water plants colonizes the streams and rivers making them to evolve slowly into dry land eventually leading to desertification.

Interferes with Fishing Activities

Some forms of solid waste contain hormones that hasten plant growth in the water bodies. For example, untreated effluent released into lakes can lead to fast growth of water weeds, e.g. water hyacinth—which interferes with the fishing activities as such weeds hinder movement of fishing vessels in the lake. Also such weeds hinder permeation of light into the water thus discouraging breeding of fish. Plastic cans and wire mesh disposed in rivers can cause damage to fishing nets thereby hampering fishing activities.

Increases Temperature of Water

Some solid wastes such as fluids and hot water are released into streams and rivers by power plants' cooling systems. The power plants use water for cooling their turbines and reactors and then discharge it into streams and rivers when it has been heated. When the temperature of water rises it speeds chemical and biological processes in rivers and streams therefore reducing the ability of water to retain oxygen dissolved in water. Consequently this increases algae growth, thus disrupting fish reproduction.

Can Lead to Extinction of Species

Most solid wastes, especially polyethylene papers and plastics, often end up in streams and rivers where they tend to wind up. In the long run these plastics, polyethylene papers and synthetic nets are carried downstream and end up in oceans and lakes. In such big water bodies they can even be seen floating in water and they are a threat to many organisms living in such water bodies. Researchers believe that plastic plays a role in rising rates of species extinction. Professor Richard Thompson from the Plymouth University in Britain quoted this, "It is evident that marine debris may be contributing to the potential for species extinction." According to reports worldwide, the researchers have found evidence that many marine organisms are swallowing debris or becoming entangled in the debris. Such entanglement has resulted in direct harm or even death of marine life. Plastic rope and netting are responsible for the majority of entanglements, whereas plastic fragments are the most ingested. Eventually most species are becoming extinct due to the intrusion of their ecosystem by solid waste pollutants.

Ways of Curbing Solid Waste Stream Pollution

Despite all these negative impacts of solid waste pollution there are various techniques by which this menace can be curbed, as discussed below.

Source reduction: Firstly, the government should come up with source reduction practices. This technique entails crafting of products in such a way that it reduces the amount of waste that will later have to be thrown away and also to make the resultant waste to be less toxic. That is minimizing the quantity of waste generated and discarded at each step of a product development or use. For instance, consumers, because they are the major contributors to solid waste additions, should decline using goods that make use of wrapping or implements made from non-biodegradable or non-compostable biodegradable materials.

Recycling factor, whereby a given solid waste material is separated from other wastes and then processed so that it can be used again in a form similar to its original use. It also involves collection of used and discarded materials and then processing them to come up with new goods. In this approach, the volume of waste that is thrown into the landfills reduces and this eventually reduces the volume of landfill space (Spiegelman and Sheehan, 2005). Recycling allows post-consumption materials to substitute virgin resources in manufacturing, therefore decreasing the need for additional trees or oil needed to produce paper products and plastics. Also solid wastes can be directly combusted to generate electrical power. The benefit is that no new fuel sources are required other than the solid waste that would otherwise be discarded in the dumping sites. Further still, the solid waste is regarded as a regeneratable source of energy because it is always available. At the electrical power generation station, the solid waste is shredded to facilitate easy handling after offloading from garbage collection trucks. Then materials that can be recycled are sorted out, while the remaining waste that remains is put in a combustion chamber for burning. The heat that is released from the solid waste that is being burnt produces steam, which in turn rotates a turbine to generate electricity.

Creating Public Awareness

The government, through the ministry of environmental conservation, and also NGOs, should from time to time hold vigorous campaigns on the solid waste streams pollution issue. They should conduct such campaigns in the affected areas so as to educate people on the dangers of polluting the rivers and streams with solid waste (Browner, Carol. 2002). By doing this the public is warned and after a period of time the people will be fully educated and become cautious. By using the mass media there should be advertisements aired on the possible dangers of stream pollution by solid waste. By practicing this, the community will be in the limelight and will assist in conserving and managing waste pollution in the streams.

Prevention of Soil Erosion

When soil is eroded by water it transfers sediments from the land and ends up in streams, rivers and other waterways. Some chemicals that exist in the soil mix with the water and therefore create complications in the life of various organisms. For example, the levels of phosphorus rises in the rivers and streams, which leads to algae blooms that lead to death of large numbers of fish. One of the most highly recommended and the best methods of preventing soil erosion is by planting a lot of trees, herbs, grasses, and fodder. This variety of plant life will help to bind the soil particles together and make them compact, hence preventing them from being carried away easily by the running water. This reduces the rate of sedimentation in the streams and rivers thus making the flow of water to be natural and appealing.

Cleaning up Stress and Rivers Through Community Based Programs

In every place where there is human recreation in or nearby rivers and streams, there is a lot of evidence of solid waste pollution. In such areas what meets the eyes are polyethylene bags, plastic containers and other forms of trash that are an eyesore to the human beings who are concerned about environmental conservation. Such kind of solid waste often ends up in rivers causing pollution. People should take bold steps by being responsible in taking care of their environment. People should pick up litter whenever they come across it. In the long run this is fairly the best and fastest way for an individual to take part in conserving the water bodies. Also, people, institutions, and even clubs should organize cleaning parties with local people so as to increase numbers and also to extend the reach (Velz, 1984). Even the participants and the general public should be encouraged through providing incentives. Also, institutions involved should get sponsors who will give prizes to the individual who collects the most litter.

Keeping of the Machinery in Good Working Conditions

The oil used to cool and lubricate engines in various kinds of machines requires to be checked and changed from time to time. If this changed old oil is not disposed in the right manner it presents a number of environmental hazards to both the human beings and the animals. As oil moves all over the crank case of an engine, it breaks down into numerous diverse types of cancer-causing and even mutagenic compounds. When an engine is dripping it discharges this poisonous oil into the roads and streets and eventually finds its way into the sewer. From the sewer it becomes part of other harmful solid wastes that end up into rivers and streams. The government should decree implementable laws targeting the automobile industry that every leaking car should be considered unroadworthy and the driver of such a car be charged by a court of law.

Reusing: In this case, people should be encouraged to embrace the idea of reusing as many things as possible before throwing them away. For example, after buying goods in a supermarket the paper bags given to carry your goods should be recycled. The same case applies to plastic containers and bottles. In many rural homes, such containers—specifically the twenty liters—are used for fetching water and putting farm produce like eggs and milk.

Finally, **composting** can be emphasized and it involves collecting organic waste, such as food scraps and trimmings from hedges and yards, and keeping it under anaerobic conditions so as to enable it to naturally decompose. After decaying, the resultant compost can then be used as a natural fertilizer. Yard waste that sits around can easily wash into storm drains when it rains. Sometimes the waste can be free of toxic chemicals and pollutants but the introduction of large quantities of trash like leaves, sticks, and grass clippings can overwhelm watercourses with unhealthy quantities of nutrients (Jenkins, J.C. 2005). People should use bins to dispose their compost to prevent the materials from being carried away by rain water. However, there are municipalities that provide them at a subsidized cost. Also, people should be exhilarated to use mulching mowers instead of collecting grass clippings and putting them in bag. This is because the mulching mowers add a natural layer of compost to the lawns and hence one does not have to deal with disposal of grass clippings.

Conclusion

Pollution as a result of solid waste streams is an issue that requires maximum attention and should be addressed by all the stakeholders involved. This is because nearly everybody is involved in solid waste pollution ranging from manufacturers all the way to consumers of the manufactured products. With the government in the forefront, incidences of solid waste pollution will reduce considerably. As years go by the amount of dumpsites will reduce and so will the negative consequences of the same. By the end of it all, we are going to live in a clean and fresh environment, thus reducing incidences of lifestyle ailments.

References

Bergeson, Lynn L. (2000), *TSCA: The toxic substances control act.* Chicago: Section of Environment, Energy, and Resources, American Bar Association.

Browner, Carol. (2002). "Polluters should have to pay for cleanups." *Chicago Daily Law Bulletin* (March 1).

Girdner, Eddie J., and Jack Smith. (2002). *Killing me softly: Toxic waste, corporate profit, and the struggle for environmental justice.* New York: Monthly Review Press.

Jenkins, J.C. (2005), *The humanure handbook: A guide to composting human manure* (3rd ed.). Grove City, PA: Joseph Jenkins, Inc.;.

Kiser, Jonathan V. L. (2003), Recycling and waste-to-energy: The ongoing compatibility success story. MSW Management.

O'Leary, P. R., & Walsh, P. H. (1995). *Decision maker's guide to solid waste management,* Vol. II. Washington, D.C.: U.S. Environmental Protection Agency.

Rogers, W. H., Jr. (1986). *Environmental law: Air and water pollution.* St. Paul, Minn.: West.

Sprankling, J. G., & Gregory, S. W. (1997). *The law of hazardous wastes and toxic substances in a nutshell.* St. Paul, Minn.: West.

Velz, C.J. (1984). *Applied stream sanitation* (2nd ed.). United States: John Wiley & Sons Inc., New York, NY.

Landfill Design for Waste Management

Jaber Almakhalas

The effect of specific practices during landfill design and operation on landfill decomposition is an area that has attracted investigation since the 1970s.

(Komilis, 1999, p. 20)

Fifteen years ago under amendments to the Resource Conservation and Recovery Act (RCRA), Congress mandated the first multilayered liner and cap systems (including geo-membranes) for hazardous waste landfills, thereby creating the basis for "dry tomb" storage of wastes.

(Richardson, 2000, p. 1)

Abstract

Waste management, being one among the foremost vital aspects of urban development, is constantly getting importance among developing nations. Landfills, that were initiated for hazardous waste management and afterward reworked into hygienic landfills, are the foremost wide custom-made practice for municipal solid waste management worldwide. However, the standard style of landfills not solely fails to fulfill the wants of waste management however conjointly fails to focus on best resource recovery and energy generation. Within the gift study, changed style was planned for part designed lowland system supported theoretical issues. Its potential for energy generation and resource exercise was analyzed with a case study of urban center municipal solid waste. It had been found that the system with changed style might yield zero.157 million tons of landfill gas (0.145 million heaps of coal equivalent) out of one year of solid waste. Further, this might recover resource valued at US$2.49 million each year.

Introduction

Rapid growth and urbanization ends up in increasing environmental issues and municipal solid waste (MSW) management is of prime

185

importance in such rising urban problems. The state of economy, to an outsized extent influences waste generation and MSW specifically. In contrast to alternative solid wastes, particularly industrial solid waste, hazardous waste, hospital waste, MSW involves waste generation from varied sources. Komilis, 1999 suggested that "Landfill design and operational practices can be used, sometimes in combination with municipal solid waste (MSW) pretreatment techniques, to control effectively landfill behavior".

The generation being the non-point/area supply, assortment and disposal poses a heavy downside to the native municipalities and alternative regulative bodies. The per capita waste generation rate among developing countries was recorded as low as three hundred g (g/capita/day) (CES, 2000). However, thanks to changes in living conditions and influence of western throw away culture, continuing raise in solid waste generation is anticipated (CES, 2000; IIT, 1997). This increase in solid waste generation not solely ends up in environmental degradation however additionally involves great loss of natural resources, that remains unaccounted for (Bagchi 2004).

MSW management involves numerous steps, specifically collection, transportation, process and disposal. Land disposal is that the most typical technique adopted. In developed countries correct landfills exist together with correct construction and maintenance of a similar (DOE/EIA, 1999). However, in developing countries properly designed and maintained landfills area unit rarely found. The waste is disposed of in open dumps that not solely results in severe environmental degradation, however conjointly leads to loss of natural resources (Parikh and Parikh, 1997). Existing designs are developed for landfills within the developed countries with the target of ultimate disposal of waste and a few provisions is formed to handle the gas generated consequently (EIIP, 1999; Johennessen, 1999). These style procedures were ne'er targeted to handle waste in integration with natural resources utilization. Even although, several landfills with gas recovery system exist within the USA, they were designed with standard methodology and therefore yield a lot of lesser landfill gas (LFG) (LMOP-USEPA, 1996).

The essential criteria for the landfill style involve the safe disposal of waste. However, method for improved rates of gas generation and its harvest are still missing. With the expected rise in waste generation within the developing countries as conjointly availableness of well-tried technology for gas recovery, a replacement and changed approach to the look of landfill system with gas recovery has been developed for the Asian conditions. Not like developed countries, Asian countries have no landfills however solely open dumps. Now, since landfills with correct style having been created obligatory by the Central Government, this new style would be a timely try. This projected design targets the new lowland sites and can't be applied to the present dumpsites. Landfills are designed to some extent to boost the gas generation rates so there may well be quicker yield of LFG and fewer land demand. LFG yield and its energy values were calculated by means that of practical models developed supported theoretical concerns.

Landfill technology

Until recently, the conception of land filling was used to dump stuff for disposal. Therefore, not abundant care was taken regarding their construction. Inserting the waste within the Earths social class was thought of because the safest follow of waste disposal. However with fast industry and urbanization, land filling has metamorphosed. As uncontrolled landfills have shown potential of polluting numerous elements of the setting and plenty of accidents have additionally happened, rules are obligatory on lowland location, web site style and their preparation and maintenance. A particular degree of engineering was created obligatory for landfills. Figure three shows the schematic illustration of landfills. Landfill location was supported several factors and an Environmental Impact Assessment (EIA) is obligatory for the siting of landfills. Land accessibility for extended marketing periods and accessibility of canopy material square measure a number of the vital guiding parameters in low frequency selected site choice.

Due to these engineering inputs, landfills are tightened up effort far better setting within the cells for better waste decomposition. As a result of careful protection of cells with daily cowl, nuisance might be reduced beside improved anaerobic conditions among the cell and also the lifetime of the lowland might be reduced. This could cause redoubled emission of LFG and its escape into the atmosphere is a potential threat to the world setting that ends up in warming. On the opposite hand, LFG has extended energy price and it depends on the composition of waste and additionally alternative conditions. Typically, the LFG collected is employed for electricity generation and boiler heating in varied industries.

Uses of Landfill technology

It was established that LFG can be used for electricity generation by utilizing each single similarly as twin fuel engines. It is used as fuel for preparation requirements and for heating boilers of varied treatment systems. Effective use of LFG in countries like the United States of America and in Europe not solely might offer resolution for waste management however conjointly contributes considerably to non-renewable energy generation and minimize GHG emission from MSW. Within the year 1997, LFG contributed as high as 952,314 x 103 kWh 3.428 x 109 MJ) of electricity within the USA and is that the next highest contributor to the renewable energy sector when electricity sector (DOE/EIA, 1999). Throughout 1993-1997, from among the biomass energy consumption within the United States of America, MSW Associate in Nursing LFG contributed a big half and it absolutely was consumed by the commercial sector to an extent of seventy- three (DOE/EIA, 1999). In the United States of America, 133 landfill sites were recovering LFG in 1997 among which around 120 tum out energy for generating facilities. These facilities have a combined generating capability of 832 MW. Most of the landfills within the USA area unit made for waste disposal solely, that ne'er thought of gas generation and harvest in their styles. Thus, the gas generation rate is lower thanks to the absence of any thought for sweetening of gas generation within the landfill designing method.

Usage in Asian countries

Though in Asian countries where land could be a tropical country, and one in every of the foremost favorable places for landfill system of waste disposal with recovery of gas, no efforts are being created for its development. High complex organic fraction and larger wetness content of Asian MSW favors gas generation significantly (Bhide, 1994). MSW in metropolis consists of concerning four-hundredth organic and complex elements with a wetness content of a lot of that fifty. In spite of all this, MSW in Asian nation is disposed of in open dumps. Because of the actual fact that MSW in Asian nation consists of a lot of organic waste and conjointly high wetness content besides prevailing tropical climate, LFG generates even in open dumps and escapes into the atmosphere. Additionally to the current, improper selling consumes a lot of space. In metropolitan cities like metropolis, considering that the prevailing dumpsites reach their full capability, waste is directed to derelict lands, that aren't in any respect appropriate for waste disposal and LFG generation and escape to the atmosphere from open dumps could be a major concern for specialists of temperature change.

Hence, to avoid this danger, the government of an Asian nation recently ordered that properly designed landfills are a compulsory demand for MSW disposal. Within the light-weight of the recently developed legislation for correctly designed landfills, MSW from metropolis with its made organic and wetness content, would end in a lot of gas generation that conjointly has to be handled. Further, because of the actual fact that paraffin could be a major constituent of LFG and has appreciable energy worth, its energy potential has to be evaluated. This might lead the means for LFG assortment from landfills and its use for varied functions.

189

Land fill gas

As waste decomposes in landfills it produces biogas with some 45% CO_2 and 55% CH_4 (Bhide, 1994; EIIP, 1999). This gas is termed as LFG. Owing to the presence of methane series, LFG incorporates a heat content of around five hundred BTU/ft3 (18629.5 KJ/m3), that is around *112* that of commercially marketed gas. Given the potential of landfills to come up with gas for energy, commitment to mitigation of GHG emissions and a high demand of land for standard landfills, there is a requirement for rising the planning and layout of landfills so all the preceding requirements are met. The projected style may even be opti-mized for economic performance.

The EPA estimates that for each a million loads of municipal solid waste, 432,000 cubic feet of landfill gas is created daily. If not controlled and monitored, the gas will migrate beneath a landfill site and cause fires and explosions. Landfills will unleash methane and greenhouse gas into the atmosphere, intensifying international warm-ing; they will additionally emit VOEs that manufacture ground-level gas or smogginess, unsafe pollutants and pernicious odors. In 1996, the independent agency ordered larger lowlands to gather and combust landfill gas by flaring it or putting in an energy recovery system. The gas should be monitored whereas a landfill is active and for 30 years once it is sealed.

Methodology

Landfill site selection

A landfill should be set and designed therefore as to meet the required conditions for preventing pollution of the soil, groundwater or surface water and making certain efficient collection of leachate. Also, a low-land website ought to be unbroken as so much as doable away from population density, for reducing pollution impact to public health. On the opposite hand, the Landfill location ought to be placed as close as doable to existing roads for saving road development, transportation, and assortment prices. What is more, the lowland website with slope either too steep or too flat is not applicable for constructing the landfill.

The landfill site should not be placed close to a residential or associate urban area, to avoid adversely affecting land price and future development and to protect the final public from doable environmental hazards free from landfill sites. The landfill should be situated inside ten km of associate urban area (Baban and Flannagan 1998) and lowland shall not be situated inside 1,000 m of associate urban area.

Waste Characteristics

Waste characteristics can give vital design data for decisions in operation procedures. Waste kind affects the handling techniques, and waste amount determines web site life, daily in operation procedures and canopy necessities. A waste characterization study ought to precede landfill siting work, however extra data is also required whereas the ability is being designed. As an example, sure waste varieties is also used as daily cover or onsite road base.

When making ready a profile of the wastes which will be received at the new landfill, concentrate to sources which will be inadvertently mixture dangerous waste with solid waste. In suspicious cases, dangerous waste testing procedures is also necessary. Systematic load checking throughout web site operation conjointly ought to be planned. The types and variety of vehicles that transport solid waste to the location ought to be tabulated, too. Traffic data are helpful for later analysis of roadways and access points.

Site layout development

The landfill's layout are greatly influenced by the site's geology. The potential for gas and leachate migration and also the suitableness of the soil for landfill base and cover material ought to be a specific concern. The location layout begins with geotechnical data, which incorporates information on the surrounding site earth science, geophysical science and soils. This information sometimes is collected throughout the location choice method, then supplemented throughout ensuing investigation.

Soil-boring logs and different information describing subterraneous formations and groundwater conditions are diagrammed to present an interpretation of the subterraneous conditions at the planned web site. Soil-boring logs facilitate to point out the extent of every formation compute between the boreholes. The depths to bedrock and also the groundwater table are also shown. More boring logs and extra cross sections at regular coordinate intervals in many (minimum of two) directions generally square measure needed to properly find the waste disposal space among the developing web site.

Phase diagrams

The site plan ought to chronologically illustrate the developing landfill's options. The landfill's end-use will begin on completed sections whereas different as within the landfill still are being employed for disposal.

Phasing diagrams show the landfill's evolution at totally different stages through the site's life. Phases ought to be developed for key times in sufficient detail to make sure that the operator is aware of what's to be done at any purpose. The engineers and management should be assured that the location is continuing in step with arrange in order that contracts is let or finances organized for construction. Restrictive bodies conjointly should be assured that landfill operators being following the plan which the locations are completed as designed at the agreed-upon time. The scale of every part area unit is determined by many factors. Generally, every part accommodates two to three years of refuse volume.

Landfill Design Steps

Determination of solid waste quantities and characteristics:

a. Existing
b. Projected or planned

Compilation of information for potential sites:

a. Performance factor for boundary and topographic surveys
b. Preparation for the base maps of existing situation on and near sites

Property boundaries, topography and slopes, surface water, wetlands, utilities roads, structures, residences, and land use.

Compilation of hydrogeological information & preparation of location map:

Soils (depth, texture, structure, bulk density, porosity, permeability, moisture, ease of excavation, stability, pH, and cat-ion exchange capacity), bedrock (depth, type, presence of fractures, and location of surface outcrops), groundwater (average depth. seasonal fluctuations, hydraulic gradient and direction of flow, rate of flow, quality, and uses).

Compilation of climatological data:

Precipitation, evaporation, temperature, number of freezing days, and wind direction.

Identification of regulations (federal, state & local) and Standards of design:

Loading rates, frequency of cover, distances to residences, roads, surface water and airports, monitoring, groundwater quality standards, roads, building codes, and contents of application for permit.

Design of filling area

a. Selection of landfilling method based on:
   Site topography, site soils, site bedrock, and site groundwater.
b. Specification design dimensions:
   Cell width, depth, length, fill depth, liner thickness. Interim cover soil thickness, and final soil cover thickness.

c. Specification of operational features:

Use of cover soil, method of cover application, need for imported soil, equipment requirements, and personnel requirements.

Design features:

a. Leachate controls
b. Gas controls
c. Surface water controls
d. Access roads
e. Special working areas
f. Special waste handling
g. Structures
h. Utilities
i. Recycling drop-off
j. Fencing
k. Lighting
l. Wash racks
m. Monitoring wells
n. Landscaping

Preparation of design package:

a. Develop preliminary site plan of fill areas.
b. Develop landfill contour plans.

Excavation plans (including benches), sequential fill plans, completed fill plans, fire, litter, vector, odor and noise controls.

c. Compute solid waste storage volume, soil requirement volumes, and site life.
d. Develop final site plan showing:

Normal fill areas, special working areas, leachate controls, gas controls, surface water controls, access roads, structures, utilities, fencing, lighting, wash racks, monitoring wells, landscaping.

e. Prepare elevation plans with cross-sections of:
Excavated fill, completed fill, phase development of fill at interim points

f. Preparation of construction details
Leachate controls, gas controls, surface water controls, access roads, structures, monitoring wells.

g. Prepare ultimate land use plan.
h. Prepare cost estimate.
i. Prepare design report.
j. Prepare environmental impact assessment.
k. Submit application and obtaining required permits.
l. Prepare operator's manual.

Conclusion

Landfill scenario has been dynamic over time. With increase in waste generation rates, scarceness of land handiness and warming problems, there is a good want for modifications to the present landfill designs targeting at energy generation from waste and with less demand of space. Manifold land saving has been attainable on account of those cells styled by the new approach as they might need solely a set space of land whereas the traditional styles would want land on an eternal basis as a result of the waste takes quite twenty years to totally get composted and settle within the case of landfill designed as per the present design procedures.

The goal of siting a landfill is to produce long environmental protection that's economically economical and complies with applicable rules. Only a few potential lowland sites are ideal. However the landfill's style part permits managers to beat web site deficiencies mistreatment verified engineering techniques. A well-developed style arrangement can create construction, operation and closure less technically tough and more cost effective.

References

Bagchi, A. (2004). *Design of Landfills and Integrated Solid Waste Management.* Wiley; NY.

Centre for Environmental Studies (CES) (2000) Down to Earth, 31 January, New Delhi, India.

Bellezza, I. (2004). Optimization of landfill volume by the simplex method. *Engineering Computations,* Vol. 21 Issue: 1, pp. 53–65

Elnokaly, A. (2013). Reducing waste to landfill in the UK: identifying impediments and critical solutions. *World Journal of Science, Technology and Sustainable Development,* Vol. 10 Issue: 2, pp. 131–142

Indian Institute of Technology (IIT) (1997) Proceedings of Workshop on Solid Waste Management and Utilization, Department of Chemical Engineering, 78 November, Mumbai, India.

Baban, S. J., & Flannagan, J. (1998). Developing and implementing GIS-assisted constraints criteria for planning landfill sites in the UK. Planning Practice and Research, 13(2), 139–151. Doi: 10.1080/02697459816157.

Gray, D. (2001). *Geotechnical Aspects of Landfill Design and Construction.* Prentice Hall; NY.

Komilis, D. (1999). *The effect of landfill design and operation practices on waste degradation behavior: a review.* doi: 10.1177/0734242X9901700104

Mncwango, S. (2005). Sanitary landfill energy harnessing and applications. *Journal of Engineering, Design and Technology,* 3(2), 127–139.

Noble, G. (1994). *Sanitary Landfill Design Handbook.* CRC Press; CA.

Richardson, G. (2001). *Coordinating Regulations in Design and Construction of Modern Landfill Liners and Closure Caps.* Retrieved from http://www.mswmanagement.com/MSW/ Articles/Coordinating_ Regulations_in_Design_ and_Constructio_4515.aspx

Rovers, F. (1994). *Solid Waste Landfill Engineering and Design.* Prentice Hall; NY.

Townsend, T. (1976). *Landfill Bioreactor Design & Operation.* CRC Press; CA.

Walsh, P., & O'Leary, P. (2002, July 1). Lesson 7: Preparing Landfill Designs & Specifications. Retrieved April 4, 2015, from http:// waste360.com/mag/wastepreparing_landfill_designs

Paper Mill Air Pollution

Emad Hawsawi

Paper mill pollution denotes the ecological contamination founded by the manufacturing, utilization and the reprocessing of paper. According to US Legal, Inc (2014), "Pulp and paper is the third largest industrial polluter to air, water, and land in the United States, and studies show that it releases well over 100 million kg of toxic pollution each year." Paper contamination causes harsh unpleasant outcomes to the eminence of air, water as well as land. Dumped papers are chief constituent of numerous landfill locations. Paper reprocessing is moreover a cause of effluence owing to the sludge generated throughout de-inking. The paper quantity along with paper products being utilized is very gigantic and the ecological outcome of paper pollution is momentous. The process of manufacturing paper requires huge quantity of clean water and generates gigantic amounts of wastewater, which is polluted with numerous macrobiotic and inorganic chemicals such as lignin, phenols and sulfides amongst others. The wastewater quantity along with attributes relies on the level of operation, the utilized raw materials as well as the engaged procedure. In general, the aim of this study is to discuss the process of paper production in terms of chemical plus mechanical pulp production and the bleaching process, together with the resultant air, water and land pollution.

In United States, pulp and paper factories are the biggest producers of pulp and paper in the world. On average, the factories manufacture about 9 million tons of pulp per year as well as "26 billion newspapers, books, and magazines" (U.S. Department of Labor, 2015). These mills are responsible for 35% of pulp generated on earth and constitute 16% of the pulp factories worldwide. Consequently, the pulp and paper factories in the United States are the cause of noteworthy quantities of contaminants, which are discharged to the atmosphere. The factories are estimated to produce about 245,000 metric tons of poisonous air contaminants per year. According to Ince, Cetecioglu and Ince (2011),

198

"Pulp and paper industry is considered as one of the most polluter industry in the world." (p. 223) However, the pulp and paper factories are a crucial employer in the United States. According to the U.S. Department of Labor (2015), the mills "are one of the nation's largest industries made up of approximately 565 manufacturing facilities located in 42 states and employ over 200,000 people."

The manufacture of papers first requires the production of pulp. The most significant element during pulp and paper production is cellulose (Bajpai, 2012, p. 1). Generally, cellulose refers to a long-chain carbohydrate made up of polymerized glucose. Cellulose creates sturdy fibres that are perfect for paper-production. To acquire the fibres of cellulose, short-chain carbohydrates known as hemicelluloses (mixtures of sugars such as glucose, galactose, and arabinose amongst others) have to be eliminated. Measured up to cellulose, the short-chain carbohydrates are effortlessly ruined and liquefied.

In addition, the woody plant substance includes a formless, extremely polymerized gist known as lignin that creates a surface layer around the fibres, cementing them as one. Lignin is as well enclosed inside the fibre. The lignin's chemistry is intricate since it is principally made up of phenyl propane components connected jointly in a three-dimensional arrangement. According to N. Agrawal, A. Agarwal and Mukherjee (2015), "The three linkages between the propane side chains and the benzene rings are broken during chemical pulping operations to free the cellulosic fibres" (p. 60). Numerous additional essences, for instance, resin acids, alcohols and turpenoid compounds are available in indigenous fibres. Nonetheless, their precise components and quantities depends on their plant resource. A majority of these elements can dissolve in water or in nonaligned solvents, and are jointly known as extractives. Pulp factories extort and process wood to obtain cellulose fibres, concurrently eliminating unnecessary ingredients like lignin.

There are three major procedures involved in pulping—namely, the mechanical, the chemical, and the semi-chemical processes. The chemical pulping process incorporates the use of three chemicals referred to as kraft, sulfite, and soda. In general, there are three paces involved during the pulping procedure. In chemical pulping, the first process involves cooking wood in a digester at a very high pressure with a liquid of the apposite elements that liquefy the lignin along with leaving the cellulose behind. This cooking procedure ends up in emanation of a diversity of dangerous air contaminants such as "methanol, formaldehyde, acetaldehyde, and methyl ethyl ketone, as well as reduced sulfur gases" (U.S. Department of Energy, 2005, p. 38).

During the mechanical pulping procedure, the wood material is compressed against a dicer, which actually detaches the fibers. This process, which is power demanding, generates an unclear product that is scrawny and loses its color straightforwardly if exposed to luminosity. The process of semi-chemical pulping utilizes an amalgamation of chemical and mechanical procedures. The chips of wood are moderately ripened with chemicals, whilst the rest of the pulping is carried out mechanically.

The second step in the pulping process involves cleansing of the pulp. Subsequent to removal of the pulp, the obtained pulp is cleansed so as to eliminate the liquefied lignin along with other chemicals. During the cleansing procedure, the pulp is passed through a sequence of cleansers and panels. The cleansing procedure takes place at elevated temperatures that produces a huge amount of exhaust fumes that contains poisonous air contaminants, which are discharged to the environment. The resultant fluid from the procedure of washing comprises lignin in addition to the chemicals employed during the separation of lignin from the cellulose. The chemical healing procedures utilized to recover the resultant chemicals results in secretions of harmful air contaminants as well.

The third pulping procedure comprises the bleaching of the pulp. Subsequent to washing, in case one desires a white product, the pulp ought to be bleached so as to eliminate the dye linked with outstanding residual lignin. The bleaching procedure also has three common approaches namely elemental chlorine bleaching, elemental chlorine free bleaching and totally chlorine free bleaching. The Elemental Chlorine Bleaching is a procedure which is presently being used at several existing bleaching factories, and utilizes chlorine (Cl) as well as twofold hypochlorite to make the pulp brighter. Additionally, Sodium hydroxide with or devoid of peroxide is employed in extracting chlorine from the pulp. According to the Central Pollution Control Board (2006), "When elemental chlorine and hypochlorite react with the lignin, they form chlorinated pollutants such as chloroform, dioxins, and furans in the wastewater stream." (p. 22)

The process of Elemental Chlorine Free Bleaching (ECF) involves replacing chlorine with chlorine dioxide as the agent of bleaching, whereby hypochlorite is never utilized. Bleaching through ECF bleaching ensures that there is a reduction in amounts of chlorinated contaminants released in the wastewater rivulet. Finally, the bleaching process of Totally Chlorine Free (TCF) does not utilize any chlorinated bleaching substance during pulp bleaching. As an alternative, bleaching substances like oxygen along with peroxide are exploited. According to the National Council for Air and Stream Improvement (2013), "Because the TCF process does not use chlorine, it does not produce chlorine residuals in wastewaters" (p. 5). In general, during the procedure of bleaching, the chemicals used to bleach are introduced into the pulp, along with the resultant combination being cleansed with water. The procedure takes place in numerous instances and produces a huge amount of liquid dissipate. In addition, ventilations from the bleaching cisterns discharge dangerous air contaminants that include chloroform, formaldehyde, and methanol amongst others.

With regards to the bleaching compounds utilized, the resultant waste torrent from the bleaching procedure could include chlorine compounds as well as organics. The assortment of chemicals might cause the development of numerous poisonous chemicals (for instance dioxins, furans, as well as chlorinated organics). Even though the runoff is in general discharged to a waste water curing plant, the toxic elements basically go through (that is, the curing plant does not succeed in reducing the pollutants concentrations) the plant where consequently they mount up in the rivers as well as seas, whereby the curing plants releases them.

There are numerous air pollution concerns that arise as a result of the paper industry. For instance, air productions of oxidized sulphur chemicals from factories of pulp and paper have resulted in devastating damages to plant life, in addition to discharges of reduced sulphur compounds that have engendered grumbles regarding "rotten egg" smells. (Aminvaziri, 2009, p. 13) Researches amongst inhabitants of pulp factory societies, particularly children, have demonstrated respiratory outcomes associated with particulate discharges, as well as the irritation of mucous membrane and headaches considered to be allied to reduced sulphur compounds. Out of the entire pulping procedures, chemical methods especially kraft pulping has the utmost prospective of causing air pollution tribulations. This is because it emits "a substantial volume of noxious and malodorous air pollutants, including particulate matter, sulfur compounds, nitrogen oxides and volatile organic compounds" (Center for Biological Diversity, 2012).

Most of the sulphite operations discharge huge amounts of sulphur oxides, particularly those utilizing calcium or magnesium bases. According to Stellman and International Labour Office (1998), "The main sources of sulphur oxides consist of batch digester blows, evaporators and liquor preparation, with washing, screening and recovery operations contributing lesser amounts" (p. 69). Kraft recuperation kilns are in addition a supply of sulphur dioxide, since they are energy boilers that exploit high-sulphur gas or oil as energy.

Hydrogen sulphide, methyl mercaptan, dimethyl sulphide and dimethyl disulphide, amongst other reduced sulphur compounds, are nearly completely allied with kraft pulping, in addition to giving these factories their distinguishing stench. These compounds mainly originate from the recovery kiln, digester blow, its reprieve valves, along with washer ventilations. Additionally, evaporators, smelt cisterns, the lime furnace as well as waste water might contribute too. A number of sulphite functions utilize reducing surroundings in their recovery heaters and might have allied reduced sulphur stench tribulations.

The gases of sulphur discharged by the recovery kiln are best proscribed by plummeting secretions at the source. They are controlled through low-odor recovery kilns, black liquor oxidation, operating the recovery furnace suitably, as well as diminution in liquor sulphidity. Sulphur fumes from digester blow, digester relief regulators along with liquor evaporation could be gathered and cremated—for instance, in the lime furnace. Incineration flue fumes may be amassed through the use of scrubbers. According to E Instruments International (2015), "Nitrogen oxides form when fuels are burned at high temperatures, as in a combustion process." Therefore, the gas may be generated in whichever factory that has a recovery boiler, power oven or a lime furnace, regarding on the operating circumstances. The generation of nitrogen oxides could be checked through the regulation of hotness, air-fuel proportions as well as time of dwelling in the incineration zone. Other gaseous chemicals are inconsequential givers to factory air contamination—chemicals such as carbon monoxide from partial incineration, chloroform as a result of bleaching functions, and unstable "organics from digester relief and liquor evaporation." (Teschke, 2008).

Particulates crop up mostly from incineration operations, although smelt-disbanding cisterns could as well be a negligible source. Over 50 percent of pulp factories particles are extremely fine (smaller than 1 μm in thickness). Such fine substance comprises sodium sulphate (Na_2SO_4) as well as sodium carbonate (Na_2CO_3) from recovery ovens, lime furnaces along with smelt-dissolving cisterns, as well as sodium chloride from smoldering by-products of woods that have been stockpiled in salt water. The discharges from lime furnace incorporate a noteworthy quantity of crude particulates owing to the calcium salts entrainment as well as sublimation of sodium elements. These might as well comprise fly ash along with products of organic incineration, particularly from power cisterns. Diminution of particle concentrations could be attained through surpassing flue fumes via electrostatic precipitators or scrubbers. Modern inventions in power cistern expertise take in liquidized bed incinerators that flame at exceedingly high hotness, bringing about more resourceful power conversion, along with facilitating smoldering of less consistent wood waste.

Pulp and paper industries also cause a lot of water pollution issues. The polluted wastewater being released from pulp and paper factories could cause the bereavement of marine organisms, permit bioaccumulation of poisonous elements in fish, as well as damaging the flavor of downstream consumption water. (Forestry Department, 2015) The emissions from pulp and paper wastewater are distinguished on the basis of corporeal, chemical or organic attributes, with the most significant being the content of solids, oxygen requirement and toxicity.

The wastewater solids substance is normally categorized on the foundation of the portion that is floating (against dissolved), the portion of floating particles that can be settled, along with the portions of either that are volatile. The portion that can settle is the most obnoxious, since it might create an opaque sludge coverlet near to the emancipation point, which swiftly reduces dissolved oxygen in the obtaining water along with permitting the propagation of anaerobic microorganisms that produce methane as well as reduced sulphur fumes. Even though non-settleable particles are typically weakened by the receiving stream and are consequently of lower anxiety, they might convey poisonous organic chemicals to marine beings. Suspended particles emitted by pulp and paper factories take in bark atoms, wood fibre, sand, gravel from mechanical pulp processing, papermaking stabilizers, fluid dregs, water curing procedure by-products along with microbial cells from minor treatment processes (Janzé, 2011).

According to Gangwar (2014), "Wood derivatives dissolved in the pulping liquors, including oligosaccharides, simple sugars, low-molecular-weight lignin derivatives, acetic acid and solubilized cellulose fibres, are the main contributors to both biological oxygen demand

(BOD) and chemical oxygen demand (COD)" (p. 3740). Chemicals that are poisonous to marine life forms incorporate chlorinated organics (AOX), resin acids, and unsaturated fatty acids, amongst others. Of meticulous concern are the chlorinated organics since they are extremely poisonous to nautical organisms along with being able to bioaccumulate. This class of chemicals, incorporating the polychlorinated dibenzo-*p*-dioxins, is the chief momentum for decreasing the exploit of chlorine in bleaching pulp.

The quantities along with supply of suspended solids, demand for oxygen as well as poisonous expulsions are dependent on procedure. Owing to the solubility of timber extractives with diminutive or without compound plus resin acid recovery, the sulphite and CTMP pulping produce intensely poisonous runoffs with elevated BOD. Kraft factories traditionally utilized a lot of chlorine for bleaching, causing their discharges to be more poisonous; nonetheless, overflows from kraft factories, which have abolished chlorine in bleaching in addition to utilizing secondary treatment normally showing modest to severe toxicity if any, as well as subacute toxicity being seriously plummeted.

Suspended solids are recently less of a predicament since majority factories exploit the primary clarification procedure, which eradicates about 95% of the settleable solids (Sappi Fine Paper North America, 2012, p. 5). There also are secondary wastewater curing technologies like aerated lagoons which are meant to diminish BOD, COD as well as chlorinated organics in the bilge water (United States Environmental Protection Agency, 2011, p. 11). Nonetheless, procedure along with operational alterations might still found sporadic toxicity breaches. A comparatively new contamination control approach to eradicate water pollution completely is the "closed mill" notion. What differentiates closed mills is that liquefied overflow is "evaporated and the condensate is treated, filtered, then reused" (Greenberg, 2003, p. 539). However, rust is a chief problem facing mills that utilize closed systems, along with concentrations of microorganisms and endotoxin increasing in cast-off process water.

Paper and pulp industry also brings about land pollution challenges. Most solids (sludges) confiscated from fluid sewage treatment systems fluctuates, in relation to their source. The primary treatment emits solids that mainly are made up of cellulose fibres. Microbial cells are the chief constituent of solids from secondary treatment procedures. In case the factory utilizes chlorinated bleaching chemicals, both primary as well as secondary solids might as well include chlorinated organic chemicals, a crucial contemplation in finding out the degree of treatment necessitated.

Before the disposal, solids are condensed in the units of gravity sedimentation where they are automatically dewatered in centrifuges or vacuum sifts. Solids from primary healing are moderately simple to remove water. According to Kay and Anya (2015), "Secondary sludges contain a large quantity of intracellular water and exist in a matrix of slime; therefore they require the addition of chemical flocculants." The moment it is adequately dewatered, slush is discarded in land-based uses such as spreading on farming or forested land, utilized as manure or burnt. Even though burning is more expensive and may add to air contamination tribulations, it could be beneficial since it can obliterate or decrease poisonous substances (for instance, chlorinated organics), which can generate staid ecological tribulations in case they were to trickle into the groundwater from other land uses.

References

Agrawal, N., Agarwal, A., & Mukherjee, S. (2015). Chemistry of raw materials for electrical grade paper: An overview. *International Journal of Engineering Technology, Management and Applied Sciences.*

Aminvaziri, B. (2009). Reduction of TRS emissions from lime kilns. Retrieved on 16th April 2015 from <https://tspace.library.utoronto.ca/bitstream/1807/18139/1/Aminvaziri_Bahar_200911_MEng_thesis.pdf>

Bajpai, P. (2012). *Biotechnology for pulp and paper processing.* Springer.; DOI 10.1007/978-1-4614-1409-4_2,

Center for Biological Diversity. (2012). EPA to review pulp mill clean air standard after 17-year delay: *Lawsuit Forces Reassessment of Outdated Emission Rules After Critics Raise Public Health, Climate Change Concerns.* Retrieved on 16th April 2015 from <http://www.biologicaldiversity.org/news/press_releases/2012/pulp-mills-08-27-2012.html>

Central Pollution Control Board. (2006). Final report on development of guidelines for water conservation in pulp and paper sector. *National Productivity Council,* New Delhi.

E Instruments International. (2015). Nitrogen oxides: What is NO_x? Retrieved on 16th April 2015 from <http://www.e-inst.com/combustion/nitrogen-oxides-nox>

Forestry Department. (2015).The EIA in the pulp and paper industry: Environmental impact assessment. *FAO.* Retrieved on 16th April 2015 from <http://www.fao.org/docrep/005/v9933e/v9933e04.htm>

Gangwar, A.K. (2014). Applicability of microbial xylanases in paper pulp bleaching: A review. *bioresources.com. 9*(2).

Greenberg, M. (2003). Occupational, industrial, and environmental toxicology. New York: Elsevier Health Sciences.

Ince, B. K., Cetecioglu, Z., & Ince, O. (2011). Pollution prevention in the pulp and paper industries.

Janzé, P. (2011). Rock removal from woody biomass. *Canadian Biomass Magazine.* Retrieved on 16th April 2015 from <http://www.canadianbiomassmagazine.ca/education/rock-removal-from-woody-biomass-3009>

Kay, T. & Anya, K. (2015). Environmental and public health issue. Retrieved on 16th April 2015 from <http://www.ilo.org/iloenc/part-x/

paper-and-pulp-industry/disease-and-injury-patterns/item/856-environmental-and-public-health-issues>

National Council for Air and Stream Improvement. (2013). Environmental footprint comparison tool: A tool for understanding environmental decisions related to the pulp and paper industry.

Sappi Fine Paper North America. (2012). Water use and treatment in the pulp and paper industry. *SAPPIPRO-5941*

Stellman, J. & International Labour Office. (1998). Encyclopaedia of occupational health and safety, Volume 3. New York: International Labour Office.

Teschke, K. (2008). Chapter 72 - Pulp and Paper Industry

U.S. Department of Energy. (2005). Energy and environmental profile of the U.S. pulp and paper industry. Retrieved on 16th April 2015 from <http://energy.gov/sites/prod/files/2013/11/f4/pulppaper_profile.pdf>

U.S. Department of Labor. (2015). Pulp, paper, and paperboard mills. Retrieved on 16th April 2015 from <https://www.osha.gov/SLTC/pulppaper/>

United States Environmental Protection Agency. (2011). Technical development document for the final effluent limitations guidelines and standards for the meat and poultry products point source category (40 CFR 432).

US Legal, Inc. (2014). Specific issues: Paper. *Environmental Law.* Retrieved on 16th April 2015 from <http://environmentallaw.uslegal.com/specific-issues/paper/>

Questions About Pollution

1. What new things did you learn about pollution from reading this chapter?

2. Explain how you think that your handling of solid waste is a proper way to prevent, or minimize pollution.
3. After reading this chapter, would you be willing to live near a land-fill? Why or why not?
4. Which types of pollution do you believe is the most harmful to the environment? Why?
5. What are some things you can do to reduce the amount of pollution in the world?

CPSIA information can be obtained at www.ICGtesting.com
Printed in the USA
LVOW10s1205181215

467145LV00002B/3/P